KEYNOTE guide to
topics in your course

Astronomy	Pages 69–88
Chemistry	33–56
Classification of Elements	41
Classification of Fossils	125
Electricity	25–32
Energy	5, 58
Gas Laws	49
Geology	89–132
Geomorphology	105–112
Heat	49
Light	21
Machines	9
Magnetism	25
Maps	97
Measurement	1
Mechanics	5
Meteorology	133–152
Minerals and Rocks	89–96
Mole Concept	49
Nuclear Physics and Chemistry	57–68
Oceanography	149
Organic Chemistry	53
Physics	1–32, 57, 68
Relativity	57
Soils	129
Solar System	69–76
Sound	17
Space Exploration	85
Stars	77–84
States of Matter	13, 68
Structural Geology	113–124
Weather	137–152

CLIFFS KEYNOTE REVIEWS

Physical Science

by

JAMES W. BATCHELOR

CLIFF'S NOTES, INC. • LINCOLN, NEBRASKA 68501

ISBN   0-8220-1732-6

© Copyright 1968

BY CLIFF'S NOTES, INC.

All rights reserved. No part of this book

may be reproduced or utilized in any form or

by any means, electronic or mechanical, including

photocopying or recording, or by any information

storage and retrieval system, without permission

in writing from the publisher.

L. C. Catalogue Card Number: 68-55296

Printed in the United States of America

CONTENTS

PHYSICS

1	Measurement	Page 1
2	Mechanics	5
3	Simple Machines	9
4	Solids, Liquids, and Gases	13
5	Sound	17
6	Light	21
7	Magnetism and Static Electricity	25
8	Electric Current	29

CHEMISTRY

9	Kinds and Properties of Substances	33
10	Practical Units and Oxidation States	37
11	The Periodic Table	41
12	Chemical Classification	45
13	Temperature, Heat, Gas Laws, and the Mole Concept	49
14	Organic Chemistry	53

NUCLEAR PHYSICS

15	Relativity and Quantum Mechanics	57
16	Radioactivity	61
17	Interconversion of Mass and Energy	65

ASTRONOMY

18	The Solar System I	69
19	The Solar System II	73
20	Stars, Galaxies, and Nebulae	77
21	The Universe	81
22	Space Exploration	85

GEOLOGY

23	Major Minerals and Rocks	89
24	Economic Minerals	93
25	Maps	97
26	Strata and Geologic Time	101
27	Geomorphology—Water and Wind	105
28	Geomorphology—Glaciers	109
29	Structural Geology	113

30	Mountains and Continents	117
31	Continents	121
32	Fossils and their Classification	125
33	Soils and Ground Water	129

METEOROLOGY

34	Atmospheric Layers and Circulation	133
35	Weather—Humidity and Clouds	137
36	Weather—Precipitation and Winds	141
37	Weather Systems	145
38	Oceanography	149
	Appendixes	153
	Final Examination	156
	Dictionary-Index	161

TO THE STUDENT

This KEYNOTE is a flexible study aid designed to help you REVIEW YOUR COURSE QUICKLY and USE YOUR TIME TO STUDY ONLY MATERIAL YOU DON'T KNOW.

FOR GENERAL REVIEW

Take the SELF-TEST on the first page of any topic and turn the page to check your answers.

Read the EXPLANATIONS of any questions you answered incorrectly.

If you are satisfied with your understanding of the material, move on to another topic.

If you feel that you need further review, read the column of BASIC FACTS. For a more detailed discussion of the material, read the column of ADDITIONAL INFORMATION.

FOR QUICK REVIEW

Read the column of BASIC FACTS for a rapid review of the essentials of a topic and then take the SELF-TEST if you want to test your understanding of the material.

ADDITIONAL HELP FOR EXAMS

Review terms in the DICTIONARY-INDEX.
Test yourself by taking the sample FINAL EXAM.

1

MEASUREMENT

SELF-TEST

DIRECTIONS. In this Self-test, and all subsequent Self-tests, write the term necessary to correctly complete the statement, or (in the case of multiple-choice questions) the letter indicating the correct term, in the answer column.

1. The system of measurement most used throughout the world is the _____ system.

2. The system whose numerical relationship is decimal is the
   a. English system
   b. metric system
   c. Latin system

3. Early measurements based on parts of the body were variable because
   a. each king changed the standard
   b. people are different sizes
   c. parts are longer in summer, shorter in winter

4. Each metric unit is _____ larger than the next smaller unit.

5. Ten centimeters is equal to
   a. ten millimeters
   b. one meter
   c. one decimeter

6. Ten centigrams is equal to
   a. one hundred milligrams
   b. one tenth of a gram
   c. one decigram

7. Three feet equals how many meters? The correct conversion factor is
   a. $0.3048 \frac{m}{ft}$
   b. $1.093 \frac{yd}{m}$
   c. $3.28 \frac{ft}{m}$

8. Three grams are how many pounds? The correct conversion factor is
   a. $0.9463 \frac{1}{qt}$
   b. $453.59 \frac{g}{lb}$
   c. $2.204 \frac{lb}{kg}$

9. The equation $x = 1/y$ is
   a. a direct proportion
   b. an inverse proportion
   c. a multiple proportion

10. The number 8.93×10^5 is equal to
    a. 893,000
    b. 8,930,000
    c. 89,300,000

11. The number 2.34×10^{-3} is equal to
    a. 0.0234
    b. 0.00234
    c. 0.000234

12. $16^6 \times 2^3$ is equal to
    a. 32^9
    b. 32^{12}
    c. 32^3

13. 60° centigrade is equal to
    a. 92°F
    b. 102°F
    c. 140°F

14. In moving a vector force line, one must keep the same
    a. orientation
    b. direction of application
    c. length
    d. all three

1 _____
2 _____
3 _____
4 _____
5 _____
6 _____
7 _____
8 _____
9 _____
10 _____
11 _____
12 _____
13 _____
14 _____

BASIC FACTS

SYSTEMS OF MEASUREMENT. Two systems of measurement are currently in wide use.

The *English system* is in use in the United States and in Canada. England itself and Australia are changing gradually to the metric system. The English system is characterized by *related units and unrelated numbers*. That is, the units form a related series such that the bottom part of one unit fraction becomes the top part of the next; but the numbers have no relationship to one another and must be memorized.

Length	Weight	Capacity
12 $\frac{in.}{ft}$	16 $\frac{oz}{lb}$	8 $\frac{oz}{pt}$
3 $\frac{ft}{yd}$	2,000 $\frac{lb}{ton}$	2 $\frac{pt}{qt}$
1760 $\frac{yd}{mi}$		4 $\frac{qt}{gal}$

(Read "12 $\frac{in.}{ft}$" as "12 inches per foot".)

The *metric system* is used commonly outside the U.S. and Canada; and internationally by scientists. It is characterized by *related units and related numbers*. That is, both the units and numbers form a related decimal series.

Length	Weight	Capacity
10 $\frac{mm}{cm}$	10 $\frac{mg}{cg}$	10 $\frac{ml}{cl}$
10 $\frac{cm}{dcm}$	10 $\frac{cg}{dcg}$	10 $\frac{cl}{dcl}$
10 $\frac{dcm}{m}$	10 $\frac{dcg}{g}$	10 $\frac{dcl}{l}$

(Read "10 $\frac{mm}{cm}$" as "10 millimeters per centimeter".) The metric system

(Continued on page 4)

ADDITIONAL INFORMATION

Most early units of measurement were based on parts of the body. These units were variable because people come in many sizes. A single country might have been standardized by royal decree, but no international standards existed. The French Academy of Science, in 1791, set international standards based on objects less variable than parts of the human body. This set of standards, called the *metric system*, has been widely accepted.

Older Variable Standards
hand—average width (approx. 4 in.)
foot—average length (approx. 12 in.)
cubit—elbow to fingertip (approx. $\frac{1}{2}$ yard)
yard—length of one arm outstretched (center of body to closed fist)
fathom—length of both arms outstretched (closed fist to closed fist)

International Metric Standard
meter—originally meant to equal one ten-millionth of the distance from the pole to the equator; now defined as 1,650,763.33 wave lengths of the orange-red line of light emitted by glowing Krypton 86
gram—mass of one cubic centimeter of water at its greatest density (4°C)
liter—volume of one kilogram of water at its greatest density (4°C)

METRIC-ENGLISH CONVERSION FACTORS.
Multiply (*Eng. units*) times: Multiply (*metric unit*) times:

2.54	$\frac{cm}{in.}$	0.3937	$\frac{in.}{cm}$
0.3048	$\frac{m}{ft}$	3.28	$\frac{ft}{m}$
0.914	$\frac{m}{yd}$	1.093	$\frac{yd}{m}$
1.609	$\frac{km}{mi}$	0.621	$\frac{mi}{km}$
453.59	$\frac{g}{lb}$	2.204	$\frac{lb}{kg}$
0.9463	$\frac{l}{qt}$	1.0567	$\frac{qt}{l}$

METRIC-ENGLISH CONVERSION STEPS.
1. Select a conversion factor that has in the bottom part of its unit fraction the same unit that you want to convert. For example, to convert 5 *mi* to the metric system, select the conversion unit factor of $\frac{km}{mi}$. Note: This is strictly a ratio problem. Do not use numbers.
2. Cancel common units. (One of them *must* be in the bottom part of a fraction.) For example, m̶i̶ × km/m̶i̶ = km

(Continued on page 4)

EXPLANATIONS

1. The metric system is used in nearly all countries except U. S. & Canada.

2. The system whose numerical relationship is decimal is the metric system, in which each unit is 10 times larger than the next smaller.

3. Early measurements based on parts of the body were variable because people are different sizes, some tall, some short; and so on.

4. This is because the metric system is based on the decimal system.

5. Ten centimeters is equal to one decimeter, since each unit is 10 times the next smaller one.

6. Ten centigrams is equal to all three.

7. If the factor is right, the unwanted unit will cancel.

$$3 \text{ ft} \times 0.3048 \frac{m}{\text{ft}} = \underline{\hspace{2cm}} m$$

8. $3 \text{ g} \times 2.204 \frac{lb}{\text{kg}} = \frac{\underline{\hspace{1cm}}}{1,000} lb$

Only these units will cancel. Since the denominator is in kilograms, divide by 1000.

9. The equation $x = 1/y$ is an inverse proportion since if y gets larger, x gets smaller.

10. Since the exponent is positive, the decimal point moves to the right (five places).

11. Since the exponent is negative, the decimal point moves to the left (three places—including integer).

12. To multiply powers, add the exponents and multiply the numbers.

13. To convert Centigrade to Fahrenheit multiply by 1.8 and add 32.

14. All three are necessary. The compass bearing shows what line the force acts along. The arrow at one end of the line shows the direction of application of the force. The length of the line, drawn to scale, shows the magnitude of the force.

Answers

metric	1
b	2
b	3
ten times	4
c	5
a,b,c	6
a	7
c	8
b	9
a	10
b	11
a	12
c	13
d	14

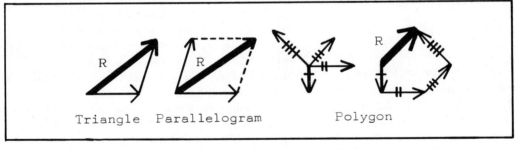

FIG. 1-1. Methods of Vector Solution.

is a *decimal system*: Each unit is 10 times smaller than the next larger unit. Each unit is 10 times larger than the next smaller unit.

Prefix Meaning   Base

		L	W	V
milli-	$\frac{1}{1000}$	meter	gram	liter
centi-	$\frac{1}{100}$	meter	gram	liter
deci-	$\frac{1}{10}$	meter	gram	liter
	1	meter	gram	liter
deka-	10	meter	gram	liter
hecto-	100	meter	gram	liter
kilo-	1000	meter	gram	liter

NUMBER-UNIT PROBLEMS.

Values for velocity, force, acceleration, and so on are found by number-unit problems. The solutions are similar to metric-English conversion problems. See Appendix I.
(a) Solve the unit step first.
(b) Solve the number step second.

PROPORTION PROBLEMS.

Direct Proportion. Like the head of a piston: the sides move similarly.

$$x = ny \text{ (n is a constant)}$$

Inverse Proportion. Like a balance scale: the sides move oppositely.

$$x = \frac{n}{y} \text{ (n is a constant)}$$

POWER AND EXPONENT PROBLEMS. An *exponent* is a small number placed at the upper right hand corner of an integer. It indicates how many times the integer is multiplied by itself. (For example, $A^3 = A \times A \times A$. A^3 is also referred to as A to the third power.)

A positive power of ten shows the number of places the decimal point moves to the right. A negative power of ten shows the number of places the decimal point moves to the left.

will cancel; but mi × mi/hr will not cancel.
3. Now, insert the numbers and multiply them.
    Ex. 5 m̶i̶ × 1.609 km/m̶i̶ = 8.645 km
    If you cannot cancel common units, you selected the wrong factor!

In *a direct proportion*, if one side of an equation goes "up," the other side goes "up" too. For example, if $x = y$, then $3x = 3y$.

In *an inverse proportion*, if one side of an equation goes "up," the other side goes "down." For example, if $x = \frac{1}{y}$, then when $y = 2$, $x = \frac{1}{2}$, or when $y = 200$, $x = \frac{1}{200}$.

Using *exponents* makes writing very small and very large numbers easier and faster. (*Note*: An exponent differs from a charge. For example, Cl^{-1} indicates a charge on a chlorine ion. This is not an exponent or a power.)

Negative powers are preceded by a minus sign; signs are not written for positive exponents.

Positive Exponents	*Negative Exponents*
$10^0 = 1$	$10^0 = 1.0$
$10^1 = 10$	$10^{-1} = 0.1$
$10^4 = 10,000$	$10^{-4} = 0.0001$
$10^6 = 1,000,000$	$10^{-6} = 0.000001$
$7.32 \times 10^6 = 7,320,000$	$7.32 \times 10^{-6} = 0.00000732$

(Read "7.32×10^6" as "seven point three two times ten to the sixth." To *multiply powers*, add the exponents. To *divide powers*, subtract the exponents. For example:
$$(2.5 \times 10^3)(6 \times 10^3) = 15. \times 10^6;$$
$$9 \times 10^6 \div 3 \times 10^2 = 3 \times 10^4.$$

Note: In writing a power, the decimal point goes between the first and second integers (numbers).

Several *rules concerning vectors* may be stated:
(1) Vector force lines may be moved to become adjacent or divergent. Divergent vector arrows start from the same point. Adjacent ones follow one another.
(2) Moved lines must keep the same orientation, direction of application, and length.
(3) The single line which represents the movement an object would have under the influence of all the forces acting upon it is called the *resultant*.
(4) To draw the resultant, *always* start back at the original point: Lift your pencil off the paper and start back at the point where you first started the drawing.
(5) A force equal and opposite to the resultant is the *equilibrant*.
(6) The three solution methods proceed as follows.
    Triangle Method: Place the vectors head to tail to form a V-shape. Draw the third side; this is the resultant.
    Parallelogram Method: Place the vectors tail to tail. Sketch the other two sides of the parallelogram. Draw the diagonal and this is the resultant.
    Polygon Method: Place the vectors tail to head (in order). The resultant closes the gap.
(See Fig. 1.1 on page 3.)

2

MECHANICS

SELF-TEST

1. An airplane flies 500 $\frac{mi}{hr}$ south. This is
   a. speed
   b. velocity
   c. acceleration

2. A car goes from 20 $\frac{ft}{sec}$ to 40 $\frac{ft}{sec}$ in 10 sec. Its acceleration is
   a. 20 $\frac{ft}{sec}$
   b. 2 $\frac{ft}{sec}$
   c. 60 $\frac{ft}{sec}$

3. In deceleration the formula is
   a. $a = \frac{v_1 - v_2}{t}$
   b. $a = \frac{v_2 - v_1}{t}$
   c. $a = \frac{v - v}{t}$

4. A communications satellite in orbit circles for six years until impact and friction with air molecules slows it and it "burns up" moving through the atmosphere. This is an example of
   a. Newton's First Law of Motion
   b. Newton's Second Law of Motion
   c. Newton's Third Law of Motion

5. A large boy and a small boy take turns pushing the same car. The large boy moves the car faster. This is an example of
   a. Newton's First Law of Motion
   b. Newton's Second Law of Motion
   c. Newton's Third Law of Motion

6. A boy pushes first a large then a small car. He moves the small one faster. This is an example of
   a. Newton's First Law of Motion
   b. Newton's Second Law of Motion
   c. Newton's Third Law of Motion

7. A jet plane takes off with a great rush of flame and exhaust gases coming out the rear. This is an example of
   a. Newton's First Law of Motion
   b. Newton's Second Law of Motion
   c. Newton's Third Law of Motion

8. The force of the sun's gravity holding the earth in orbit is an example of
   a. centripetal force
   b. centrifugal force
   c. an unbalanced force

9. A child rides at constant speed on a merry-go-round. His acceleration is
   a. constant
   b. increasing
   c. nonexistent

10. The gravitational attraction of the earth holds a 100 lb rock on its surface. The gravitational attraction of the rock for the earth is
    a. none
    b. less than that of the earth's for the rock
    c. equal to that of the earth's for the rock

11. A bomb is dropped from an airplane. Its acceleration is
    a. continuously increasing
    b. continuously decreasing
    c. constant

12. In the Law of Universal Gravitation, the relationship of force to distance is
    a. direct proportion
    b. equal
    c. inverse proportion

1 _____
2 _____
3 _____
4 _____
5 _____
6 _____
7 _____
8 _____
9 _____
10 _____
11 _____
12 _____

BASIC FACTS

FORCE, MOTION, AND GRAVITY.
Uniform Motion.

Speed = $\dfrac{\text{distance}}{\text{time}}$, or

$s = \dfrac{d}{t} \left(\dfrac{mi}{hr}, \dfrac{ft}{sec}, \dfrac{m}{sec}, \text{etc.} \right)$

$v = \dfrac{d}{t}$ (Where v stands for velocity, when the direction is given)

Accelerated Motion.
Acceleration (rate of change in velocity):

$a = \dfrac{v_2 - v_1}{t} \left(\dfrac{ft/sec}{sec} \text{ or } \dfrac{ft}{sec^2} \right)$

Deceleration:

$a = \dfrac{v_1 - v_2}{t} \left(\dfrac{ft/sec}{sec} \text{ or } \dfrac{ft}{sec^2} \right)$

v_1 occurs before v_2 in time. When using acceleration or deceleration formulas, *always write the "v" with the largest value first.*

NEWTON'S LAWS OF MOTION.
First Law: Objects at rest remain at rest, and objects in motion remain in motion, until acted on by an outside force.
Second Law. Acceleration is directly proportional to force, and inversely proportional to mass.
Third Law. For each action, there is an equal and opposite reaction. This law is an expression of the conservation of momentum (which is the product of an object's mass times its velocity).

Circular Motion.
  Centripetal Force—IN:

$$F_c = \dfrac{mv^2}{r}$$

  Centrifugal Force—OUT:

$$-F_c = -\dfrac{mv^2}{r}$$

Note: Circular motion is accelerated motion, even at constant speed.

(Continued on page 8)

ADDITIONAL INFORMATION

Speed-Velocity Example: Velocity indicates direction of movement, speed does not. For example, 40 $\dfrac{mi}{hr}$ is speed, 40 $\dfrac{mi}{hr}$ North is velocity.

DON'T MEMORIZE THE FORMULAS— LEARN THEIR PARTS BY ASSOCIATION.

Acceleration Formula: Since acceleration is a change in velocity, two velocities are shown. Since acceleration is a rate, the velocities are divided by time.

Acceleration Example. By definition, the formula for acceleration will have to be $a = \dfrac{v - v}{t}$. The hardest part then is deciding which has the largest value and comes first, v_1 or v_2. Think of the motion of a car: during acceleration (stepping on the gas), v_2 is larger; during deceleration (stepping on the brake), v_1 is larger.

Units for Acceleration: Since acceleration is velocity divided by time, the units for acceleration will be velocity units divided by time units. Thus, $\dfrac{ft}{sec}$ divided by sec will be an acceleration unit. We have, then, $\dfrac{ft/sec}{sec}$. Invert and multiply. This becomes $\dfrac{ft}{sec} \times \dfrac{1}{sec} = \dfrac{ft}{sec^2}$.

First Law: This law is an expression of inertial resistance to change. Examples of inertia would be a heavy table staying in place until pushed or pulled; or heavy ball rolling until friction or impact slows or stops it.

Newton's Second Law: Symbolically, Newton's second law could be written $a = \dfrac{F}{m}$ (acceleration is equal to force divided by mass), or $F = ma$ (force is equal to mass times acceleration).

Third Law Examples:
Gases move backwards ← → balloon moves forward.
Push against wall → ← wall resists push. This may be expressed as:

$$\text{mass}_1 \times \text{velocity}_1 = \text{mass}_2 \times \text{velocity}_2$$

For example:
  large rifle × little kick = small bullet × high velocity

Circular Motion Examples: Centripetal Force holds an object in orbit (like the force of the sun's gravity does). It converts velocity from straight line (inertia) motion to circular orbital motion.

Centrifugal Force is the equal and opposite force to centripetal force. It tends, for example, to move the earth out of

(Continued on page 8)

Mechanics 7

EXPLANATIONS

1. This is velocity because the direction of the movement is indicated.

2. $a = \dfrac{v_2 - v_1}{t} = 40\,\dfrac{\text{ft}}{\text{sec}} - 20\,\dfrac{\text{ft}}{\text{sec}}$

    $= \dfrac{20\,\frac{\text{ft}}{\text{sec}}}{10\,\text{sec}}$ (invert and multiply)

    $= 2\,\dfrac{\text{ft}}{\text{sec}^2}$

3. In deceleration the formula is $a = \dfrac{v_1 - v_2}{t}$. The velocity with the largest absolute value always comes first. In deceleration, v_1 (before applying the brake) is larger than v_2.

4. The "burning up" of a radio satellite as it moves through the atmosphere is an example of Newton's First Law of Motion: The satellite tends to continue in motion until outside force (air-molecule impact & friction) act upon it.

5. This is an example of Newton's Second Law of Motion, $a = \dfrac{F}{m}$. A larger force (push) produces a larger acceleration. (Acceleration is directly proportional to force: $a = F \cdot \dfrac{1}{m} \cdot$)

6. This is an example of Newton's Second Law of Motion (mass is inversely proportional to acceleration: $a = \dfrac{1}{m} \cdot F$).

7. A jet plane's forward movement from rearward expulsion of exhaust gases is an example of Newton's Third Law of Motion. The action and the reaction are equal and opposite. The unbalanced forward force moves the plane forward.

8. This is an example of centripetal force, a force which is directed toward the center of a circle.

9. The direction, and therefore the velocity, change. A change in velocity is acceleration. Since the speed is constant, the acceleration is also constant.

10. Gravitation applies equally between any two objects. The rock has little appreciable effect on the earth though.

11. Its velocity increases with time ($v = gt$); but its acceleration (rate of change of velocity) is constant $\left(g = 32\,\dfrac{\text{ft/sec}}{\text{sec}}\right.$ or $\left.32\,\dfrac{\text{ft}}{\text{sec}^2}\right)$.

12. In the Law of Universal Gravitation, the relationship of force to distance is an inverse proportion:

    $$F = G \cdot \dfrac{m_1 m_2}{d^2} \qquad F \propto \dfrac{1}{d^2}$$

    (The symbol, $\propto$, means "is proportional to".)

Answers

b	1
b	2
a	3
a	4
b	5
b	6
c	7
a	8
a	9
c	10
c	11
c	12

8 Mechanics

Work, Power, and Energy

Terms		Units	Formulas	Definitions
Force	(F)	lb (newtons)	$F = ma$	*Force* is a "push" or "pull" which can change the state of rest or motion.
Work	(W)	ft-lb (newton-meters)	$W = Fd(\cos A)$*	*Work* is the product of a force times the distance over which the force acts.
Power	(P)	ft-lb/sec (newton-m/sec)	$P = \dfrac{Fd}{t}$	*Power* is the rate of doing work (work divided by time).
Energy	(E)	ft-lb (newton-meters)	$E = Fd$	*Energy* is the ability to do work. Note that it uses the units of work.

Note the orderly progression of units from force to work to power. See *Appendix II* for systems of units.

*A is the angle between F and D; if they are parallel, $\cos A = 1$.

Acceleration Due to Gravity. The formula for a (velocity divided by time) may be written $v = at$. If the direction of the velocity caused by gravity is downward, $v = gt$, where $g = 32 \dfrac{\text{ft}}{\text{sec}^2}$ or $9.8 \dfrac{\text{m}}{\text{sec}^2}$. Weight equals mass times the acceleration due to gravity ($w = mg$).

Law of Universal Gravitation

$$F = G\frac{m_1 m_2}{d^2}$$

Law of Conservation of Energy: Energy can neither be created nor destroyed, but may be transformed from one form to another.

Law of Conservation of Mass: Mass can neither be created nor destroyed, but may be transformed from one form to another.

orbit, or to overturn a car on curve. Note that in circular motion, even if the speed is constant, the direction and, hence, the velocity is continuously changing. A change in velocity is acceleration.

In the acceleration due to gravity, velocity (v) increases as the time of fall (t) increases, but the rate of increase (g) is constant. If v is downward, $F = mg$ because $w = mg$. Remember that weight (w) varies with location, while mass (m) doesn't. The values for g are sea-level figures. They will be slightly lower on a high mountain top and decrease slowly going away from the center of the earth.

Any two *objects* in the universe, *attract* each other with a force that is directly proportional to the square of their masses and inversely proportional to the square of the distance between them. Gravity is a force of attraction only, not repulsion. Since division is by d^2, not just d, F decreases rapidly as an object moves out away from the earth.

KINDS OF ENERGY. Heat, light, sound, chemical, electrical, magnetic, gravitational, nuclear, and mechanical energy are all forms of energy. Mechanical energy may be either potential or kinetic.

Potential energy is the energy of position or deformation.

$$E_p = \text{weight} \times \text{height} = wd$$

Kinetic energy is the energy of movement.

$$E_k = \tfrac{1}{2} \text{ mass} \times \text{velocity}^2 = \tfrac{1}{2} mv^2$$

(Note that kinetic energy is more dependent on velocity than on mass because v is squared).

3

SIMPLE MACHINES

SELF-TEST

1. Does the formula that follows show IMA (Ideal Mechanical Advantage) or AMA (Actual Mechanical Advantage)?

   $$_____ = \frac{\text{force distance}}{\text{load distance}}$$

2. Ideal Mechanical Advantage ignores the effects of _____.

3. Write the proper operation symbol (+, −, ×, ÷) in the simple machine formula; force _____ force distance = load _____ load distance.

4. The speed of a machine is _____ proportional to its IMA.

5. AMA would equal IMA in a(an) _____ machine.

6. The _____ of a machine equals:

   $$\frac{\text{"useful" work output}}{\text{work input}} \times 100\%$$

7. The combination force, load, fulcrum identifies a _____ class lever.

8. The combination force, fulcrum, load identifies a _____ class lever.

9. The combination fulcrum, force, load identifies a _____ class lever.

10. In a movable pulley, with the weight supported on two strands of rope, the IMA is _____.

11. IMA = $\frac{\text{large radius}}{\text{small radius}}$ applies to the _____.

12. A 50 lb barrel is rolled up a board 10 ft long to a truck bed 4 ft off the ground. The force required is _____ lb.

13. A chisel is an example of a(an) _____.

14. In the screw, IMA = circumference ÷ _____.

15. A jeep traveling at 60 mi/hr approaches a hill. As it shifts to second gear its IMA doubles; in first, it triples. Its speed in second is reduced to _____ mi/hr; in first, it is reduced to _____ mi/hr.

16. A machine produces 80 ft-lb of work. However, because of friction, it requires 200 ft-lb input to accomplish this. What is the machine's efficiency?

17. A 20 ft long seesaw (1st class lever) is supported at the middle. A 40 lb girl sits at one end. How far out from the center of the seesaw must a 50 lb boy sit to balance the board?

18. What force would be needed to raise a 200 lb load 15 ft using a pulley, if the free end of the rope moved 30 ft?

19. What force would be needed to lift a 200 lb weight using a wheel and axle with a 3 in. axle and a 2 ft crank? What is its IMA?

20. With a lever arm 20 in. long, a house jack has a circumference of turning of just a little over 125 in. The distance between threads is $\frac{1}{4}$ in. What is the IMA?

1 _____
2 _____
3 _____
4 _____
5 _____
6 _____
7 _____
8 _____
9 _____
10 _____
11 _____
12 _____
13 _____
14 _____
15 _____
16 _____
17 _____
18 _____
19 _____
20 _____

BASIC FACTS

SIMPLE MACHINES. There are six basic machines. All machinery is built of multiple combinations of these.

Mechanical Advantage (MA) is the ratio of load to force. There are two cases of mechanical advantage:

Ideal Mechanical Advantage

$$IMA = \frac{\text{force (effort) distance}}{\text{load (resistance) distance}}$$

Actual Mechanical Advantage

$$AMA = \frac{\text{load}}{\text{force}} = \frac{\text{resistance}}{\text{effort}}$$

Simple Machine Formula.

force × force distance
   = load × load distance

The speed of a machine is inversely proportional to its ideal mechanical advantage.

The efficiency of a machine may be expressed as:

$$Eff = \frac{\text{``useful'' work output}}{\text{work input}} \times 100\% = \frac{AMA}{IMA} \times 100\%$$

(Power may replace the "useful" work output term.)

The lever is a rigid bar that pivots on a fixed point. The fixed point is called the *fulcrum*, and is symbolized by a triangle.

There are three *classes of levers* (based on the geometric location of parts):

F△L: fulcrum between force and load
FL△: load between force and fulcrum
△FL: force between fulcrum and load

A seesaw is an example of a F△L lever; a nut-cracker is an example of a FL△ lever; and a pair of tweezers is an example of a △FL lever.

The machine formula applies to levers.

F△L (first class) levers are more *efficient* than FL△ (second class) or △FL (third class) levers.

(Continued on page 12)

ADDITIONAL INFORMATION

The six basic machines are the lever, the pulley, the wheel and axle, the inclined plane, the wedge, and the screw. These are the primitive machines that were known and used in ancient times.

Machines are very useful because they enable us to move heavy loads with only small applications of force. This ratio of load to force is the *mechanical advantage* (MA) of the machine. *Ideal mechanical advantage* ignores the effects of friction and is used for working simple problems. *Actual mechanical advantage* includes the effects of friction and is more complicated. Notice that IMA deals only with the distances the load or force moves.

The *simple machine formula* is the basis for working most of the problems concerning machines. It is a simple proportion and usually three of the four factors are known.

Since the speed of a machine is inversely proportional to IMA, low speeds are of greater mechanical advantage than high speeds. This is why a car is shifted to low gear in mud or snow. In contrast to this, a low IMA allows for greater speed.

In a *frictionless machine*, AMA would equal IMA and work output would equal work input. However, there are no frictionless machines; therefore *efficiency*, when expressed as a percent, is always less than 100% or, when expressed as the ratio of AMA to IMA, less than one. In most elementary problems, however, friction is ignored unless the contrary is stated. This book will follow that procedure.

Use of a *lever* involves the application of a force at one point in order to move a load at another point. The machine formula applies to the lever thusly: force × force distance (distance from force to fulcrum) = load × load distance (distance from load to fulcrum).

The *efficiency* of a lever is determined by the arrangement of force, load, and fulcrum, and by the force and load distances.

In a pulley, the load is divided equally among the supporting strands. The IMA equals the number of strands, and the force needed to raise the load equals the weight on any one strand. Thus, increasing the number of strands raises the IMA and lowers the force needed.

(Continued on page 12)

EXPLANATIONS

1. The formula is for *IMA* because, in *IMA*, friction can be disregarded by using the distances instead of forces.

2. This is, of course, ideal (or theoretical) and not practical.

3. The simple machine formula, by definition, is: force × force distance = load × load distance.

4. A machine under load is shifted into low gear to get a greater IMA, with more power but less speed. Under a lesser load, the machine is shifted into high gear, resulting in a smaller IMA, with less power but higher speed.

5. In actual practice, all machines lose energy, mainly through friction-generated heat which is dissipated to the atmosphere.

6. By definition, $\frac{\text{"useful" work output}}{\text{work input}} \times 100\%$ = the efficiency of a machine.

7. A second class lever has a higher IMA than a third class lever.

8. Always use the highest class of lever possible to move a load with the least effort.

9. A third class lever is least efficient.

10. In a movable pulley, the IMA is the same as the number of strands the weight is supported on.

11. The key to this is the word *radius*. The formula must apply either to the pulley or to the wheel and axle. The number of strands of rope are not mentioned so it is not a pulley. The large radius of the crank (wheel) gives the wheel and axle its advantage. The simple machine formula applies here: force × FORCE DISTANCE = LOAD × load distance.

12. force × force distance
    = load × load distance

$$F \times 10 \text{ ft} = 50 \text{ lb} \times 4 \text{ ft}$$
$$F = \frac{200 \text{ ft-lb}}{10 \text{ ft}} = 20 \text{ lb}$$

13. One edge of a chisel tapers to a fine edge, and the opposite edge is thick. The chisel is a splitting wedge.

14. The distance between adjacent threads, or the pitch, is inversely proportional to the IMA. The smaller the pitch (finer threads), the greater the IMA.

15. The IMA of a machine is inversely proportional to its speed. When the IMA is multiplied by 2, the speed becomes multiplied by $\frac{1}{2}$.

$$60 \text{ mi/hr} \times \frac{1}{2} = 30 \text{ mi/hr}$$

When the IMA becomes multiplied by 3, the speed becomes multiplied by $\frac{1}{3}$.

$$60 \text{ mi/hr} \times \frac{1}{3} = 20 \text{ mi/hr}$$

16. $\frac{80 \text{ ft-lb}}{200 \text{ ft-lb}} \times 100\% = \frac{8000 \text{ ft-lb} \cdot \%}{200 \text{ ft-lb}}$
    = 40% efficiency

17. force × force distance
    = load × load distance
$$40 \text{ lb} \times 10 \text{ ft} = 50 \text{ lb} \times d$$
$$d = \frac{40 \text{ lb} \times 10 \text{ ft}}{50 \text{ lb}}$$
$$d = \frac{400 \text{ ft-lb}}{50 \text{ lb}} = 8 \text{ ft}$$

18. $F \times 30 \text{ ft} = 200 \text{ lb} \times 15 \text{ ft}$
$$F = \frac{3000 \text{ ft-lb}}{30 \text{ ft}} = 100 \text{ lb}$$

19. $F \times 2 \text{ ft} = 200 \text{ lb} \times 0.25 \text{ ft}$
$$F = \frac{50 \text{ ft-lb}}{2 \text{ ft}} = 25 \text{ lb}$$
$$\text{IMA} = \frac{R}{r} = \frac{2 \text{ ft}}{0.25 \text{ ft}} = 8$$

20. IMA = $\frac{\text{circumference of turning}}{\text{pitch}}$
    = $\frac{125 \text{ in.}}{0.25 \text{ in.}} = 500$

Answers	
IMA	1
friction	2
×	3
inversely	4
frictionless	5
efficiency	6
second	7
first	8
third	9
two	10
wheel and axle	11
20 lb	12
wedge	13
pitch	14
30, 20	15
40%	16
8 ft	17
100 lb	18
25 lb, 8	19
500	20

12 Simple Machines

The pulley is a wheel with a grooved rim which holds the rope or cable supporting the load.

The *IMA* of a pulley equals the number of strands of rope or cable which support the load.

There are three *classifications* of pulleys: fixed (1 strand), movable (2 strands), and combination (4 or more strands).

The *machine formula* applies to pulleys.

The wheel and axle consists of a wheel (or crank) attached to a smaller wheel (or axle).

The *IMA* of a wheel and axle is the ratio of the radius of the crank to the radius of the axle.

$$IMA = \frac{\text{large radius}}{\text{small radius}} = \frac{R}{r}$$

The *machine formula* applies to the wheel and axle.

The inclined plane is taken to be a tilted plane surface.

$$IMA = \frac{\text{force (actual) distance}}{\text{load (effective) distance}}$$

The *machine formula* applies to the inclined plane.

The wedge is a pair of inclined planes placed back to back.

$$IMA = \frac{\text{length (l)}}{\text{thickness (t)}}$$

The *machine formula* applies to wedges.

The Screw is an inclined plane wound spirally around a central core.

There are four types of screws: the holding (wood) screw, the lifting screw (car jack), the propelling screw (turbine blade), and the measuring screw (micrometer).

$$IMA = \frac{\text{circumference}}{\text{pitch}}$$
$$= \frac{\text{force distance}}{\text{load distance}}$$

The *machine formula* applies to the screw.

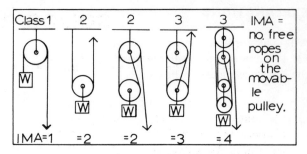

Fig. 3.1. Pulleys.

The machine formula applied to pulleys thusly: force × force distance (distance free end of rope moves) = load × load distance (distance load moves).

In a *wheel and axle* machine, force is applied to the larger wheel, and the weight (load) is attached to the smaller wheel. For the wheel and axle, force × *distance force moves* = *load* × distance load moves.

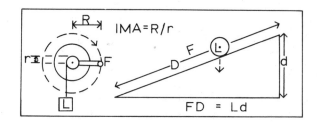

Fig. 3.2. Wheel and Axle; Inclined Plane.

The *force distance* is measured along the tilted plane surface. The *load distance* is the vertical height the object is raised.

Nails, needles, and pins are examples of *intruding wedges*. Other sorts of intruding wedges may be driven under a load in order to raise it. Axes and chisels are *splitting wedges*; and razors and knives are *cutting wedges*.

Use of the *machine formula* for the wedge involves determining the *force distance* which is measured along the inclined plane, and the *load distance* which is the height of the thick end (or the diameter) of the wedge.

The spiral surface of the inclined plane of a *screw* is called the *thread*. The distance, parallel to the central core, between two adjacent threads is called the *pitch*. The circumference of the circle of turning (the distance moved by the force) is the numerator of the *IMA*. The circle of turning may be increased (as in a jackscrew) by providing a long handle (thus increasing the radius). The *machine formula* for the screw is: force × distance force acts (n × circumference) = load × distance load acts (vertical height or n × pitch), where n is the number of complete turns.

4 SOLIDS, LIQUIDS, AND GASES

SELF-TEST

1. The state of matter in which the matter assumes the shape of its container but maintains its value is called the _liquid_.

2. A particle of matter that cannot be chemically subdivided into simpler substances is called a(an) _atom_.

3. Molecules are composed of _atoms_.

4. Atomic theory explaining chemical compounds in terms of atoms and molecules was set forth in the 1800's by _Dalton_.

5. The state of matter in which the molecules have the highest level of energy is the _gas_ state.

6. Disregarding differences between isotopes, atoms of a particular element are all _alike_.

7. Atoms combine in small whole _numbers_.

8. Molecules of gases are constantly _in motion_.

9. The behavior described in question 8 is shown experimentally by the phenomenon called _Brownian_ movement.

10. A substance may go from a gas to a liquid when energy is _removed_.

11. Force per unit area is _stress_.

12. Deformation resulting from stress is _strain_.

13. Atoms (or molecules) joined to form regular external shapes bounded by smooth plane faces are called _crystals_.

14. Solids having no regular external shape are called _amorphous_.

15. A substance that deforms slowly and never quite returns to its original shape when the deforming force is removed is called _plastic_.

16. Resistance of liquids to internal flow is _viscosity_.

17. Sea level atmospheric pressure in $\frac{lb}{in.^2}$ is _14.7 lbs/sq in_.

18. Fluid pressure at a given depth = height × _____.

19. The specific gravity of water is _1.000_.

20. The units of specific gravity are _none_.

21. The changing of matter from a liquid to a gas is called _vaporization_.

22. In a closed container, the number of molecules leaving a liquid surface and the number returning to it reach a condition of equilibrium where the vapor is said to be _saturated_.

23. Vaporization only at the surface of a liquid is called _evaporation_.

24. When vaporization takes place within a liquid, as well as at its surface, _boiling_ is said to take place.

1 _____
2 _____
3 _____
4 _____
5 _____
6 _____
7 _____
8 _____
9 _____
10 _____
11 _____
12 _____
13 _____
14 _____
15 _____
16 _____
17 _____
18 _____
19 _____
20 _____
21 _____
22 _____
23 _____
24 _____

Solids, Liquids, and Gases

BASIC FACTS

STATES OF MATTER. The three most common states of matter are the *solid*, *liquid*, and *gaseous* states. A fourth, *plasma*, is relatively uncommon at normal conditions (see Chapter 17). Recently a fifth (extremely dense) supermetallic state of matter has been postulated.

ATOMIC THEORY. Matter consists of various *elements*. The smallest observable units of these elements are *molecules* which are composed of *atoms*.

In 1808, John *Dalton* proposed an atomic theory to explain chemical compounds in terms of atoms and molecules. Dalton's theory has been revised during the years since 1808.

MOLECULAR THEORY. Molecular theory explains the behavior of solids, liquids, and gases in terms of the motions and energy-levels of molecules. The molecules of solids have the least energy and, thus, least motion. Excluding plasma, the molecules of gases have the most energy and motion.

Motion of molecules was experimentally shown by the *Brownian movement*.

The states of matter are *energy-levels*. Adding energy may change a solid to a liquid or a gas; while extracting energy may change a gas to a liquid or a solid.

FORMS OF SOLIDS. Solids may assume a *crystalline* form: common example of crystals are ice, sugar, salt, diamond, and most chemicals and minerals. Solids may also assume an *amorphous* form: for example, glass, plastic, and rubber.

Solids may be: malleable, elastic, ductile, plastic, sectile, or brittle.

Stress and Strain. Stress is the force on a substance per unit area.

(Continued on page 16)

ADDITIONAL INFORMATION

Solids retain their shape and resist deformation. *Liquids* assume the shape of their container (they flow) and resist compression. *Gases* fill their container (they disperse) and are easily compressed. *Plasma*, which exists only at extremely high temperatures, will be discussed in Chapter 17 under nuclear physics.

Bits of matter that cannot be divided into simpler substances in the ordinary chemical laboratory are called *elements*. The smallest particle of an element or compound that can exist in a free state by itself is the *molecule*. The smallest particle of an element that can take part in a chemical reaction is the *atom*.

Dalton's atomic theory contains five important theses:
1. Every element consists of tiny particles called atoms.
2. Atoms of a particular element are all alike. (This is held to be true today, except for isotopes.)
3. Atoms cannot be *chemically* subdivided.
4. Chemical combinations of elements (compounds) are formed by atoms joining together to make molecules.
5. Atoms combine in small whole numbers. (This is not true of some organic compounds.)

The central thesis of the *molecular (kinetic) theory* is that molecules of matter are constantly in motion.

The molecules of *solids* do not have enough kinetic energy to break the bonds between molecules, and, therefore, they are held in place. Their only movement is rotation or vibration around a fixed point.

The molecules of *liquids* have enough energy to break these intramolecular bonds partially. Thus, they move freely while still maintaining "contact" with one another. The motion of the molecules of a liquid is translational, i.e., from one place to another.

The molecules of *gases* have enough energy to break the molecular bonds completely. Thus, they move freely and with no mutual contact except collision. The molecules of a gas move randomly and continually in all directions.

Brownian movement is observed in the behavior of extremely small particles suspended in a fluid (a liquid or a gas). Such particles exhibit an irregular motion which is interpreted to mean that they are randomly bombarded on all sides by molecules.

A *crystalline solid* is comprised of atoms or molecules forming a well-defined external geometric shape with smooth plane surfaces. The atoms or molecules of the crystal have an orderly internal arrangement.

(Continued on page 16)

EXPLANATIONS

1. Although both gases and liquids assume the shape of their container, only the gas will expand if unconfined.

2. *Chemically* means through techniques used in the ordinary chemical laboratory, as contrasted with the techniques of nuclear physics.

3. Molecules are composed of atoms.

4. John Dalton, an English school teacher, set forth an atomic theory in the early 1800's.

5. The molecules of a gas have more kinetic energy, and therefore more motion, than those of liquids and solids. Molecules of solids have the least energy and motion.

6. By definition of the term element, the atoms of any particular element are all alike.

7. This is fortunate, for it would be much more difficult to deal with $H_{1762}O_{691}$ instead of H_2O.

8. Molecules of gases have enough kinetic energy to overcome the attraction between molecules.

9. Brownian movement is a continual erratic movement of very small particles under the random bombardment of molecules.

10. When there is less energy present, more molecules become entangled in other molecules' fields of force, and this facilitates the forging of molecular bonds.

11. Stress and pressure are both defined as force per unit area.

12. Strain is a change in dimension resulting from stress.

13. The regular external shape of the crystal is an outward reflection of an orderly internal arrangement of atoms or molecules.

14. The Greek prefix *a* means without, *morphous* is derived from the Greek word meaning form. Thus, amorphous means without (crystalline) form.

15. If a solid returns to its original shape when the deforming force is removed, it is elastic; if it returns only partly, it is plastic.

16. Viscosity is defined as internal resistance to liquid flow.

17. Sea level atmospheric pressure is $14.7 \frac{lb}{in.^2}$, equivalent to 760 mm of mercury.

18. By definition, fluid pressure = height × the density of the fluid.

19. Water is the standard for the relative scale of specific gravities, its specific gravity being set at one. The specific gravities of all other substances are determined by comparison with water.

20. Specific gravity is a ratio — a pure number which has no units.

21. Vapor is another term for gas. Therefore, the change of matter from liquid to gas is called vaporization.

22. When a vapor is saturated, it has all the molecules of the liquid it can hold (in equilibrium) at that particular temperature. If a molecule enters the gas, one must leave it.

23. Evaporation produces a cooling of the liquid as the molecules with the most kinetic energy escape, leaving the slow molecules behind.

24. Boiling is characterized by bubbles, turbulent flow, and convection cell movement.

Answers

liquid state	1
element	2
atoms	3
Dalton	4
gaseous	5
alike	6
numbers	7
in motion	8
Brownian	9
removed	10
stress	11
strain	12
crystals	13
amorphous	14
plastic	15
viscosity	16
14.7	17
density	18
one	19
none	20
vaporization	21
saturated	22
evaporation	23
boiling	24

Strain is the deformation resulting from stress. Strain is either a distortion or a dilation. *Distortion* is a change in shape. *Dilation* is a change in volume without a change in shape. Distortion is the result of unbalanced stress. Dilation is the result of balanced stress.

Percent strain = $\dfrac{\text{change in dimension}}{\text{original dimension}} \times 100\%$

The *stress-strain diagram* relates stress and strain to the behavior of different types of solids.

Deformation and Rupture. Prior to rupture (breakage), solids deform in two stages, bending (or folding) and flowage.

PROPERTIES OF LIQUIDS. Liquids flow readily but do not rupture. They exhibit *viscosity* (internal resistance to flow), and therefore *pressure* (force per unit area) is easily transmitted.

ATMOSPHERIC PRESSURE. *One atmosphere* of pressure is defined as the pressure at sea level at 45°N latitude and 0°C. This quantity is:

14.7 $\dfrac{\text{lb}}{\text{in.}^2}$ = 1013.2 millibars

DENSITY (d) is defined as mass per unit volume. *Weight-density* (D) is defined as weight per unit volume. D = d × acceleration due to gravity. D = dg

Specific Gravity. The specific gravity of a substance is the ratio of the density of that substance to the density of a standard substance (usually water).

An *amorphous solid* is comprised of a relatively haphazard arrangement of atoms. Amorphous solids are relatively uncommon in nature.

A *malleable solid* is one which can be hammered or rolled into flat sheets. A *ductile solid* is one which can be drawn out into fine wires. A *sectile solid* can be cut with a knife. An *elastic solid* deforms easily and returns to its original shape when the deforming force is removed. A *plastic solid* deforms slowly and never quite returns to its original shape. A *brittle solid* shatters easily.

Types of Rupture of Solids

	Separational movement		Frictional movement
	Uneven break	Clean break	
Mineral scale	fracture	cleavage	shear
Formation scale	fissure	joint	fault

Frictional movement is a lateral movement and leaves either a planed-off surface or a crushed zone.

Archimedes' Principle of Fluid Buoyancy states that a body immersed in a fluid is buoyed upward by a force equal to the weight of the fluid it displaces. (Apparent weight loss = weight displaced).

Three *laws of fluid pressure* are especially important:

1. The pressure at any certain depth-level in a fluid column is equal to the product of the depth (height) of the fluid multiplied by its weight-density.

2. If a force exerts pressure on the surface of a column of fluid, the force per unit area is transmitted equally throughout the fluid.

3. At a given depth-level in a fluid column, pressure is equal in all directions.

One atmosphere of pressure is equal to a barometric height reading of 76 cm, 760 mm, or 29.92 inches. *One millibar* is equal to 0.001 bar or one million dynes per square centimeter.

Density is expressed in grams per cubic centimeter. *Weight-density* is expressed in pounds per cubic foot. Because *specific gravity* is a ratio, it has no units. Some common substances, their densities, and their weight-densities are:

Ethyl Alcohol (20°C)	d = 0.79	D = 49.4
Water (4°C)	d = 1.00	D = 62.4
Gasoline	d = 0.68	D = 42
Mercury	d = 13.6	D = 850

In the cgs system, *density* is numerically equal to *specific gravity*. In the fss (English) system, weight density is equal to specific gravity multiplied by the weight density of water. The *specific gravity of water* is, by definition, 1. If a substance has a specific gravity less than 1, it floats in water.

5

SOUND

SELF-TEST

1. An energy wave transmitted only by the motion of the particles of a material medium is _____ .

2. An energy wave that travels fastest in a vacuum is _____ .

3. The speed of sound in air at 0°C, in $\frac{ft}{sec}$, is _____ .

4. The wave in which particle motion is parallel to wave motion is called _____ .

5. The wave in which particle motion is at right angles to wave motion is called _____ .

6. The distance between corresponding points on a wave is called its _____ .

7. Identical waves with one wavelength path difference between them produce _____ .

8. Identical waves with one-half wavelength path difference between them produce _____ .

9. The lowest tone of the overtone series is called the _____ .

10. A scale of unequal intervals is the _____ .

11. An "unpleasant" sound produced when the intervals will not reduce to small whole numbers is called _____ .

12. Places of maximum particle movement are called _____ .

13. To lower the pitch of a wind instrument, the air column must be _____ .

14. The speed of sound in air at 20° C, in $\frac{ft}{sec}$, is _____ .

15. The maximum displacement of a sound wave particle (in one direction) from the center line is called the _____ .

16. A sound wave moves at 1100 ft/sec from an "A" tuning fork which has a vibration frequency of 440 vibrations/sec. What is its wavelength?

17. Identical waves separated by any distances other than a whole number or half-number of wavelengths will _____ cancel or reinforce each other.

1 _____
2 _____
3 _____
4 _____
5 _____
6 _____
7 _____
8 _____
9 _____
10 _____
11 _____
12 _____
13 _____
14 _____
15 _____
16 _____
17 _____

Sound

BASIC FACTS

SOUND is an energy wave transmitted by the vibratory motion of particles.

Sound waves travel only in a material medium.

Sound waves are capable of exciting the auditory nerve, and thus can be heard.

THE SPEED OF SOUND in air is 1087 feet per second, or 331 meters per second, at 0° C. The speed of sound changes by 2 feet per second, or 0.6 meters per second, for each 1° C change in temperature.

TYPES OF SOUND WAVES. Two types of energy waves transmit sound: *longitudinal waves* in which the particles vibrate parallel to the direction of wave propagation, and *transverse waves* in which the particles vibrate at right angles to the direction of propagation.

Sound waves are usually represented by the sine curve of trigonometry.

PROPERTIES OF SOUND WAVES. The *wave length* (λ) is the distance between two successive areas of maximum or minimum particle movement (peaks or troughs). (See Fig. 5.1 on page 19.)

The *amplitude* (a) is the maximum displacement of a particle.

The *frequency* (f) is the number of complete wavelengths passing a given point in one second.

The *period* (T) is the time it takes one wave length to pass a given point.

$$\lambda = \frac{\text{speed}}{f} \qquad T = \frac{1}{f}$$

The *interval* between two waves is the lowest ratio between their vibration frequencies. For example:

$$\text{interval} = \frac{\text{high } f}{\text{low } f}$$

(Continued on page 20)

ADDITIONAL INFORMATION

Sound waves are initiated by the vibrations of some object (for example, a bell which has been struck). These vibrations create a disturbance which causes particles in the surrounding medium (for example, air) to vibrate. These particles, however, do not travel. Rather, the disturbance travels; and the sound wave consists of propagation of this energy to successive "layers" of particles.

Sound waves differ from electromagnetic waves (light, radio, radar, etc.) because sound waves must travel in a material medium while electromagnetic waves travel best in a vacuum.

The *speed of sound* varies directly with the density of the medium in which the sound wave is transmitted. Sound travels fastest in the most dense mediums, and more slowly in less dense mediums.

Sound waves are longitudinal in gases and liquids, but transverse in solids. Transverse waves are absorbed by, and therefore move only limitedly in, liquids. All electromagnetic waves are transverse.

When a source of sound has the same *frequency* as a second (potential) source, the objects are said to be in *resonance* with each other. When a source of sound is coupled with a resonator, the sound is reinforced (made louder).

Total *reinforcement*, or *constructive interference*, occurs when two waves which are of the same wavelength have their crests superimposed. Two waves are said to be *in phase* if they begin and end together for each wave length. Reinforcement results in a louder and fuller sound; or, in the visible part of the electromagnetic spectrum, a brighter light.

Extinction, or *destructive interference*, occurs when two waves of the same wavelength have crests and troughs superimposed. Two waves are said to be of *opposite phase* when one of them begins as the other is half-completed, i.e., when they differ by one-half a wavelength. Cancellation is complete if the two waves have the same amplitudes. Extinction results in the absence of sound or, in the case of a light wave, light.

Partial reinforcement or cancellation (extinction), or *partial interference*, occurs when two waves having any other wavelength relationship are superimposed. In this case, the waves alternately reinforce and cancel each other, resulting in *beats*. The resultant sound periodically rises and falls in intensity.

To summarize interference phenomena: Waves that are a whole number of identical wavelengths apart reinforce. Waves that are a half-number of identical wavelengths apart extinguish. Waves that are any other distance apart partially interfere.

(Continued on page 20)

EXPLANATIONS

1. Sound is transmitted only by a material particle transmitting its vibration to another particle.

2. The electromagnetic wave travels best when there is nothing there to hinder it.

3. The speed of sound is $1087 \frac{ft}{sec}$ at $0°C$ at $45°N$ latitude. It changes $2 \frac{ft}{sec}$ for each $1°C$ temperature change.

4. Longitudinal wave:

5. Transverse wave:

6. Wavelength is measured between corresponding points, usually crests.

7. Identical waves that are separated by a whole number of wavelengths reinforce each other, causing stronger light or louder and fuller sound.

8. Identical waves that are separated by one-half or a multiple of one-half wavelengths extinguish each other and give no light or sound.

9. The fundamental tone is the lowest or base tone. The overtones of a series have frequencies which are whole-number multiples of fundamental tones.

10. The intervals of the diatonic scale are not equal.

11. An interval that reduces to $\frac{69}{37}$ is an example of a sound that would be considered discordant.

12. The node is a place of no particle movement, so the opposite (the place of maximum particle movement) is called the antinode.

13. To lower pitch, lengthen the air column. Remember this: the slide-trombone player shoves the slide out when he reaches for a low note.

14. $1087 \frac{ft}{sec} + \left(2 \frac{ft/sec}{°C} \times 20°C\right)$
$= 1087 \frac{ft}{sec} + 40 \frac{ft}{sec} = 1127 \frac{ft}{sec}$ at $20°$

(If the temperature had gone *down* $20°$, the 40 ft/sec would be subtracted.)

15. Note that in determining the amplitude you do not add together the displacement from both sides.

16. The wave length from an "A" tuning fork $1100 \frac{ft}{sec}$ is:

$$\lambda = \frac{speed}{f} = \frac{1100 \frac{ft}{sec}}{440/sec} = 2.5 \text{ ft}$$

17. Identical waves that are any distance apart in their path difference other than a whole number or half-number of wavelengths will give partial cancellation or reinforcement.

Answers	
sound	1
electromagnetic	2
1087	3
longitudinal	4
transverse	5
wavelength	6
reinforcement	7
extinction	8
fundamental	9
diatonic	10
discordant	11
antinodes	12
lengthened	13
$1127 \frac{ft}{sec}$	14
amplitude	15
2.5 ft	16
partially	17

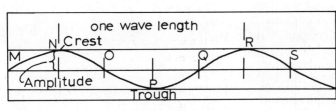

FIG. 5.1. Sine Wave.

If the frequencies are 300 cycles and 240 cycles, then the interval = $\frac{5}{4}$.

INTERFERENCE RELATIONSHIPS BETWEEN TWO WAVES. Two identical sound waves moving in the same direction will *reinforce* each other if they are *in phase*. They will *extinguish* each other if they are of *opposite phase*. And they will *partially reinforce*, or *partially cancel*, each other if they are *out of phase* or *are not identical*.

AUDIBLE FACTORS OF SOUND. A *tone series* is comprised of a *fundamental tone* and its *overtones*. *Distinguishing factors* of sound, such as *pitch*, *loudness*, and *quality*, help tell one tone from another.

The *pleasure-factors* in sound are harmony and discord. *Harmony* is the positive (pleasurable) factor, and *discord* is negative.

MUSICAL SCALES. Two commonly used scales are the *diatonic scale* in which there are 8 notes to an octave, and in which the intervals between the notes are unequal; and the *equally-tempered scale* in which there are 13 notes, at equal intervals, in each octave.

BEATS AND BEAT TONES. *Beats* are periodic reinforcements resulting from partial interference. A *beat tone* is a new tone which occurs when the beats exceed 16 per second.

TYPES OF MUSICAL INSTRUMENTS. *Wind* instruments include the trumpet, trombone, clarinet, oboe, and organ. *String* instruments include the violin, viola, cello, and piano. *Percussion* instruments include the tamborine, chimes, xylophone, and drums.

CHANGE OF PITCH. Pitch may be varied by changing the length of the vibrating part of the source of sound. If the length is increased, the pitch is lowered. If the length is decreased, the pitch is raised.

VOCAL SOUND. In *speech*, the *consonant* is the most important sound. It starts and stops the audible tone. In *singing*, the *vowel* is more important.

A *fundamental tone* is the lowest tone of a tone series emitted by a vibrating musical instrument. An *overtone* is a tone whose frequency is a whole-number multiple of the fundamental tone.

Pitch, the relative position of a tone in the musical scale, is directly proportional to frequency. *Loudness*, the intensity of a sound, depends on the amplitude of the sound waves. *Quality* enables the ear to distinguish a tone from others of the same pitch and intensity.

Sounds *harmonize* when their intervals reduce to small whole numbers. They are *discordant* when the intervals will not reduce to small whole numbers.

Because of its unequal intervals, the *diatonic scale* is not suitable for most musical instruments. Instead, the *equally-tempered scale* is used. In this scale, the frequency of each note is 1.06 times that of the note preceding it. The eight white keys of an octave on a piano constitute a diatonic scale. The eight white keys and the five black keys of such an octave together constitute an equally-tempered scale. In Oriental countries, scales other than the diatonic or equally-tempered scales are used.

In the United States, musical instruments are usually tuned to the "American" or "Philharmonic" pitch where the "A" tone is 440 vibrations per second. Tuning instruments in this way eliminates unwanted *beats* and *beat tones*. Two untuned instruments can produce *3* tones!

In *wind instruments*, a vibrating air column sets up standing waves. *Standing waves* are identical waves moving in opposite directions. At the *nodes* of a standing wave, there is no particle motion. At the *antinodes*, there is maximum particle motion. In *string instruments*, vibrating strings are coupled with a resonance box. In *percussion instruments*, the whole instrument or a part of it vibrates.

In some instruments (for example, the piano and the organ), *pitch is varied* through the use of different parts of various lengths. In other instruments, however, the effective length of one part is varied. In some cases (for example, the clarinet and oboe), this is accomplished by opening or closing holes in the side of the instrument. In other cases (for example, the trumpet and trombone), valves or slides change the passage length. Finger-pressure is used to shorten the length of strings in such instruments as the violin.

6

LIGHT

SELF-TEST

1. Light is
   a. a sound wave
   (b) an electromagnetic wave
   c. a seismic wave

2. Light is
   a. a particle
   b. a wave
   (c) both a particle and a wave

3. Electromagnetic waves travel best in a(an) ~~straight line~~ *vacuum*

4. Radar waves have a wavelength that is *longer* than that of light.

5. One ten-millionth of a millimeter is
   a. a quantum unit
   b. a spectrum unit
   (c) an Angstrom unit

6. The foot-candle is the number of
   (a) lumens/sq ft
   b. lux/sq m
   c. dial/sq ft

7. The ratio of the speeds of light in two different media is the
   a. Law of Reflection
   b. Law of Refraction
   (c) Index of Refraction

8. An image that can be shown on a screen is a(an) ~~virtual~~ *real* image.

9. A concave lens is
   (a) thicker in the middle
   b. the same throughout
   (c) thinner in the middle

10. A convex lens *magnifies* ~~inverts~~ the image.

11. A refracting telescope uses *lenses*.

12. The retina of the eye resembles the camera
    a. shutter
    (b) film
    c. diaphragm

13. Solid objects produce color by
    a. addition
    (b) subtraction
    c. division

14. Nearsightedness must be corrected by
    a. converging lenses
    (b) diverging lenses
    c. contact lenses

15. The iris of the eye resembles the camera
    (a) shutter
    b. film
    (c) diaphragm

16. The eye being too long from front to back causes
    a. nearsightedness
    b. farsightedness

17. The eye being too short from front to back causes
    a. nearsightedness
    b. farsightedness

18. Lights produce color by
    (a) addition
    b. subtraction
    c. division

1 _____
2 _____
3 _____
4 _____
5 _____
6 _____
7 _____
8 _____
9 _____
10 _____
11 _____
12 _____
13 _____
14 _____
15 _____
16 _____
17 _____
18 _____

BASIC FACTS

The Dual Nature of Light. Light behaves both as a particle (P) and as a wave (W).

Phenomena	Behavior of Light
Straight-line motion	P W
Refraction	P W
Reflection	P W
Interference	W
Diffraction	W

Electromagnetic Waves. Most of the energy waves with which physics is concerned (sound excluded) are electromagnetic waves, which travel fastest in a vacuum. Electromagnetic waves propagate with the speed of light (c), which is 186,000 $\frac{mi}{sec}$ in a vacuum.

Photometric Measurement.

Property	Unit
Luminous intensity	standard candle
Luminous flux	lumen
Illumination	foot-candle or lux (meter-candle)

$$Illumination = \frac{candle\ power}{distance^2} = \frac{I}{d^2}$$

(Where d is the distance to the light source.)

Photometric Instruments. The *photometer* compares the luminous intensity of two sources of light.

The *foot-candle meter*, or "light-meter," measures the illumination at a specific point.

Different *amounts of illumination* should be used for different purposes.

Reflection and Refraction. Whenever a light wave, or light ray, strikes a surface, it is partially *reflected* and partially *refracted*.

Reflection is *regular* when the reflecting surface is smooth, and *diffuse* when the reflecting surface is irregular.

Refraction is bending.

(Continued on page 24)

ADDITIONAL INFORMATION

Seventeenth-century scientists were in disagreement concerning the nature of light. *Newton* regarded light as a stream of particles, while *Huygens* treated it as a wave motion. Today, a combination of these views is accepted. Light is thought to be emitted as groups of particles which behave as a wave. Each such particle is called a *quantum* or a *photon*, and a group of such particles is called a *quanta*. Light (and other electromagnetic radiation) is emitted and absorbed *only in whole numbers of quanta*. The energy of each quantum is inversely proportional to its wavelength.

Electromagnetic disturbances are the simultaneous emission of both electrical and magnetic waves. These *electromagnetic waves* are *transverse waves* which travel at the *speed of light*.

The Wavelengths and Uses of Some Electromagnetic Waves

Commercial radio	600 m – 200 m
Shortwave radio	50 m – 10 m
Television	6 m – 0.1 m
Radar	1 cm – 1 mm
Infrared light	25,000 Å – 7900 Å
Visible light	7900 Å – 4000 Å
Ultraviolet light	4000 Å – 1000 Å
X-rays	100 Å – 0.1 Å
Gamma rays	less than 0.1 Å

(An *angstrom unit* (Å) is equal to 10^{-7} mm.)

The color of light is determined by it wavelength. Lights of different color may interact to form other colors. Objects appear colored because they reflect light of a certain color (the "color" of the object), while they absorb all other light.

Luminous intensity is the amount of light emitted in all directions from a point source, and is measured in terms of *candle power* (cp). *Luminous flux* is the amount of light energy emitted in one specific direction by the source in a unit of time. Luminous flux is measured in terms of *lumens*. One lumen is equivalent to 0.00147 watt, and 4π lumens are equivalent to one *cp*. *Illumination* is luminous flux per unit area, and is measured in terms of the *foot-candle* or the *lux*. One foot-candle is one lumen per square foot, one lux, or one meter-candle, is one lumen per square meter.

In the *photometer* two light sources illuminate a translucent screen situated between them. In the center of the screen is a spot of wax which cannot be distinguished from the rest of the screen when the illumination from both sources is the same. At equilibrium, the intensity of one of the sources can be calculated, if the intensity of the other is

(Continued on page 24)

EXPLANATIONS

1. Light is only one very small portion of the electromagnetic spectrum. It travels best in a vacuum (unlike sound).

2. Light is now considered to be both a particle and a wave. Particles of light (called quanta or photons) emerge in bundles or packets (not singly). These packets then travel with a wave motion.

3. Electromagnetic waves travel best when there is nothing to impede them.

4. Radar waves, TV waves, and radio waves are all of longer wavelength than light waves.

5. The Angstrom unit is one ten-millionth of a millimeter or one hundred-millionth of a centimeter.

6. Lumens/sq ft is the English system unit; lux/sq m is the metric system unit.

7. One of the two different media is usually air, so for all practical purposes the Index of Refraction is the ratio of the speed of light in a medium compared to the speed of light in air. When this ratio is high, a spreading of wavelengths and a brilliant display of colors results, as in the case of diamond.

8. In a real image, all the light rays from the lens or mirror pass through the image, so it can be shown on a screen. In a virtual image they do not pass through, so it can not be shown.

9. The term *concave* means "with cavity."

10. The convex lens is thicker in the middle and thinner at the edges.

11. Refracting means bending. The lenses bend light rays. Mirrors can only reflect.

12. Both the retina of the eye and the film of the camera register the effects of light and produce chemical reactions from it.

13. The color of a solid object is the color that it reflects. In other words it is the color that is left after it subtracts all other colors by absorption.

14. A nearsighted eye converges the light rays in front of the retina, since the eye is too long or the lens too powerful. A diverging (concave) lens is needed to spread the rays and focus them farther back.

15. The iris of the eye resembles the camera diaphragm because the function of both is to control the amount of light let in.

16. In farsightedness, close objects are focused behind the retina. Only far objects focus on the retina.

18. Colored lights produce their color effects by the addition or combining of colors (contrary to the subtraction of solid objects and pigments). What we see are the colors produced by this merging.

Answers

b	1
c	2
vacuum	3
longer	4
c	5
a	6
c	7
real	8
c	9
magnifies	10
lenses	11
b	12
b	13
b	14
c	15
a	16
b	17
a	18

In *regular reflection* and in *refraction*, the incident ray, the reflected or refracted ray, and the normal to the surface all lie in the same plane. In *diffuse reflection*, light is scattered.

The *index of refraction* (n_r) is the ratio of the speeds of light in two different media.

$$n_r = \frac{\text{speed in } m_1}{\text{speed in } m_2}$$

OPTICAL AIDS. *Lenses, mirrors,* and *prisms* are optical aids.

A *convex lens*, which is thicker in the middle than at the edges, converges light to a point and magnifies an image. The *focus* is the point at which the light rays converge. The *focal length* is the distance between the focus and the center of the lens (or other optical aid).

A *concave lens*, which is thinner in the middle than at the edges, spreads light and reduces an image.

A *reflecting telescope* uses a *mirror* to enlarge an image. A *refracting telescope* uses a *lens*. Binoculars use matched *prisms*.

IMAGES. Optical aids reflect or refract light rays, and so form images.

A *real image* is one through which the light rays actually pass. It can be seen by looking into the optical aid, or it can be produced on a screen placed where the image is.

A *virtual image* is one through which light rays only appear to pass, but through which they do not actually pass. A virtual image can be seen *only* by looking into the aid.

THE HUMAN EYE. Light is admitted to the human eye by opening or closing the *eyelid*. The *iris* controls the amount of light admitted. The *lens* focuses on a surface, and the *retina* registers it.

Simple eye defects can be corrected by the use of optical aids.

known, by using the formula: $\frac{I}{d^2}$ (first source) = $\frac{I}{d^2}$ (second source).

In the "*light meter*," a photo cell converts light falling on it to an electric current which is registered as foot-candles.

Recommended Illumination Levels in foot-candles: General living, 10; Classroom Work, 30; Detailed Work, 50; Displays and Sports areas, 100 or more.

The *law of reflection* states that the angle at which the light ray strikes a surface (the angle of incidence) is equal to the angle at which it is reflected from the surface (the angle of reflection). This law applies in the case of regular reflection.

The *law of refraction* states that when light goes from a less dense to a more dense medium, the ray is bent toward the perpendicular ("normal") to the boundary; and when light goes from a more dense to a less dense medium, the ray is best away from the normal.

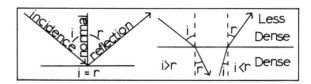

FIG. 6.1.

In an *absolute index of refraction*, m_1 is a vacuum, and the speed of light in m_1 is equal to c. In a *relative index of refraction*, m_1 is some material medium, usually air. When the index of refraction is high, a colorful display of dispersed light is observed (as in a diamond). The relative index is usually used, since the difference between it and the absolute index is very slight.

The eye retains an image for about one-fifteenth of a second. A series of still pictures run at 16 or more frames per second is seen as a continuous moving picture.

Nearsightedness (myopia) is an eye defect which impedes one's ability to see objects at a distance. It can be corrected by using eyeglasses with *concave* (diverging) lenses. *Farsightedness* (hypermetropia) impedes one's ability to see objects which are close to the eye. It can be corrected by using eyeglasses with *convex* (converging) lenses.

7 MAGNETISM AND STATIC ELECTRICITY

SELF-TEST

1. "Antimagnetic" substances are called _____.

2. Ferromagnetic substances have many electron spins that are _____.

3. The name of a paramagnetic alloy of aluminum, nickel, and cobalt is _____.

4. The most strongly diamagnetic substance is _____.

5. Points of concentration of magnetic force are called magnetic _____.

6. Lines of magnetic force are conventionally drawn from _____.

7. Unlike magnetic poles _____.

8. A pole strength that will exert a force of one dyne on an equal pole 1 cm distant in air is called a(an) _____.

9. A vertically pivoted compass used to prospect for metallic magnetic ore deposits is called a(an) _____.

10. A positive charge on an atom results from electron _____.

11. A charged atom is called a(an) _____.

12. The charge of the electron is _____.

13. The position of the proton in the atom is _____.

14. The charge of the neutron is _____.

15. The creation of an electric charge in a neutral object by placing it near a charged one is called electrostatic _____.

16. A Leyden jar is a type of _____.

17. Like electrical poles _____.

18. Magnetism and electricity are apparently related to _____.

19. The difference between magnetic and geographic north is the magnetic _____.

20. The differences between magnetic and geographic north over large areas are shown by _____ lines.

21. Magnetism and electricity are effective on a(an) _____ scale.

1	_____
2	_____
3	_____
4	_____
5	_____
6	_____
7	_____
8	_____
9	_____
10	_____
11	_____
12	_____
13	_____
14	_____
15	_____
16	_____
17	_____
18	_____
19	_____
20	_____
21	_____

BASIC FACTS

FUNDAMENTAL FORCES. There are four fundamental forces in the universe: magnetism, electricity, gravity, and nuclear force.

TYPES OF MAGNETIC SUBSTANCES. *Ferromagnetic* substances are strongly and conspicuously magnetic. *Paramagnetic* substances are weakly magnetic. *Diamagnetic* substances are "antimagnetic"—they are repelled by magnets.

ORIGIN OF MAGNETISM. Two situations create magnetism:
Unmatched electron spins in atoms create magnetism in metals.

The *transfer of electrons* from one atom to another in an electric current flowing in a wire creates a circular magnetic field around the wire.

Representation of Magnetism.
Each magnet has two *poles*, a *north* pole (N) and a *south* pole (S). The poles of a magnet *represent* the points at which force is concentrated.

Law of Magnetic Action. Like poles repel each other, and *unlike poles* attract each other.

←NN→   ←SS→   N→←S

A *magnetic field* is *represented* graphically by *lines of force* conventionally drawn from the north pole to the south pole.

MAGNETIC FORCE LAW. The magnitude of a *magnetic force* (F) between two poles a distance d apart is given by:

$$F = k \frac{P_1 P_2}{d^2}$$

where P_1 and P_2 are the strengths of the poles.

A *unit pole* is one which repels an exactly similar pole at a distance of 1 cm with a force of 1 dyne, and an exactly similar pole at a distance of 1 m with a force of 1 newton.

(Continued on page 28)

ADDITIONAL INFORMATION

Ferromagnetic substances include iron, steel, nickel, and cobalt. Permalloy (iron and nickel) and Perminvar (iron, nickel, and cobalt) are examples of ferromagnetic alloys.

Among *paramagnetic* substances are aluminum, manganese, oxygen, platinum, and many metallic salts. Paramagnetic alloys include Alnico (aluminum, nickel, and cobalt) and Silmanal (silicon, manganese, and aluminum).

All substances which are neither ferromagnetic nor paramagnetic are *diamagnetic*. Bismuth is strongly "antimagnetic." Copper, zinc, silver, gold, lead, and some types of glass are also diamagnetic.

THEORIES OF ORIGIN OF MAGNETISM BY SPIN DIRECTION. Since each piece of a bar magnet cut in half develops its own north and south poles, magnetism is a property of the smallest particle of matter. First groups of molecules and then groups of atoms were thought to be involved. Finally, the *directions of electron spins* were considered. Normally, the directions of spin in electron-pairs "cancel" each other; and, therefore, these substances tend to be diamagnetic. Ferromagnetic and paramagnetic substances, however, have uncancelled spins in the 3d incomplete electron subshells of their atoms; and the effect of these uncancelled spins is not overcome by the atoms' diamagnetic tendencies. Heating or beating a magnetic substance may change the directions (alignment) of the electron spins and so demagnetize the substance (make diamagnetic).

Most compasses are magnetized steel bars with "north-seeking" and "south-seeking" *poles*. If a compass is placed in the vicinity of a magnet, and moved in the direction to which its N pole points, it will trace the *lines of force* of the magnet. Also, iron filings sprinkled in the vicinity of a magnet will form a pattern which shows the lines of force.

The *magnetic poles of the earth* do not coincide with the geographic poles. Magnetic north differs from the geographic north by an angle called the *declination*, which is shown by *isogonic lines*.

A *vertically pivoted compass*, or dip needle, is used in prospecting for magnetic ores. An *electromagnet* is used to lift heavy magnetic materials, and in electric circuits. A *magnetometer* measures the changes in the magnetic properties of rock formations.

(Continued on page 28)

EXPLANATIONS

1. Diamagnetism opposes ordinary magnetism.

2. The many unmatched electron spin directions of ferromagnetic substances cause its magnetic effects. Diamagnetic substances have matched or paired electron spins.

3. ALuminium + NIckel + CObalt = ALNICO.

4. Bismuth wire will turn across a magnetic field rather than line up with it.

5. Lines of magnetic force appear to converge toward and diverge away from magnetic poles.

6. Conventionally, diagrams show magnetic lines of force going from N to S poles. The line traces the path a free N pole would take if it could exist in the field.

7. *Attraction* occurs between unlike magnetic poles.

8. The term "unit" implies the numerical factor one. Note that everything in the definition of the unit pole carries this through.

9. The vertically pivoted compass needle should point horizontally at the magnetic equator, straight down at the magnetic poles, and dip anywhere between, hence it is called a *dip needle*.

10. If a negative electron is removed, a positive proton in the nucleus is unmatched.

11. An ion is an atom that has gained or lost electrons.

12. The term electron is always taken to be the negative electron.

13. The proton is a nuclear particle.

14. The neutron has no charge; it may be considered a proton combined with an electron.

15. A charged object "induces" a charge in an adjacent neutral one.

16. A condenser is a device to store an electric charge. The Leyden jar is one type of condenser.

17. Like poles always repel, whether they are electrostatic poles or electromagnetic poles.

18. Magnetism, electricity, and gravity are related. Write the formulas for finding the force of each and see.

19. Magnetic declination is the angle (shown graphically for that location on topographic maps) between the magnetic geographic north poles.

20. Isogonic lines are lines of equal magnetic declination. For example: every point on the 8° E isogonic line has a magnetic declination of magnetic north being 8° east of geographic north.

21. Magnetism and electricity are most effective on an atomic or submicroscopic scale.

Answers

diamagnetic	1
unmatched	2
Alnico	3
bismuth	4
poles	5
N to S poles	6
attract	7
unit pole	8
dip needle	9
loss	10
ion	11
negative	12
in the nucleus	13
nonexistent	14
induction	15
condenser	16
repel	17
gravity	18
declination	19
isogonic	20
atomic	21

MAGNETIC FIELD OF THE EARTH. The *earth* has a variable *magnetic field* which changes considerably over long periods of time.

KINDS OF ELECTRIC CHARGES. *Electric charge* may be positive or negative. An *electron* has a negative charge. A *proton* has a positive charge.

A charged atom is called an *ion*. An atom becomes positively charged if it loses an electron, and negatively charged if it gains an electron.

LAW OF STATIC ACTION. *Like* charges repel each other, and *unlike charges* attract each other. The region in the vicinity of a charged body is an *electric field* which is represented by *lines of force*.

ELECTROSTATIC FORCE LAW. *Coulomb's law of force* between charges, q_1 and q_2, separated by a distance d, is $F = k \frac{q_1 q_2}{d^2}$.

A *unit charge* is one which will exert a force of 1 dyne on an equal charge at a distance of 1 cm.

The *electrostatic series* is a list of substances each of which becomes positively charged if rubbed against a substance below it in the series, and negatively charged if rubbed against a substance above it in the series.

CONDUCTORS AND CONDENSERS. Substances in which electrons can move freely are *conductors* of electricity. Substances which are nonconductors are called *insulators*. Layers of conductors interspersed with insulators form *condensers* (*capacitors*) which store charges.

INDUCTION. Magnetism and electric charge can be *induced* in neutral (nonmagnetized or noncharged) objects.

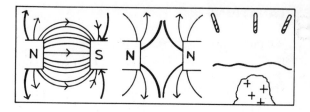

FIG. 7.1.

The two kinds of electric charge were called *positive* and *negative* by Benjamin Franklin, who thought electricity was an excess or a deficiency in an invisible weightless fluid.

The three most fundamental constituents of the atom are the *proton* (+), the *electron* (−), and the *neutron* (no charge). Atoms are neutral (have no charge) when the number of electrons is equal to the number of protons. An *electric charge on an atom* usually results from the loss or gain of an electron, because the orbiting particles are displaced easily while a great deal of force is required to displace a nuclear particle.

Lines of force represent the paths along which force is greatest in an *electric field*. They are conventionally drawn (from top to bottom) consists of: fur, flannel, polished glass pole) to the point of concentration of negative charge (negative pole). Lines of force and poles are convenient visual aids, but have no physical existence.

Coulomb's law of force is similar to the law of gravitational force, $F = k \frac{m_1 m_2}{d^2}$, and the magnetic force law, $F = k \frac{P_1 P_2}{d^2}$. The force of gravity is perceptible only when large masses are involved. Magnetism and electricity, however, are effective on the atomic scale. Gravity is only an attractive, never a repulsive, force.

Static electricity, as opposed to current electricity, is concerned with charges which are, for the most part, at rest. It is encountered when long hair is combed, when a deep pile rug is walked upon, and so forth. The electrostatic series (from top to bottom) consists of: fur, flannel, polished glass, silk, shellac, sealing wax, ground glass, hard rubber, and the metals. A hard rubber rod, for example, becomes negatively charged when rubbed against fur.

Three types of *condensers* are: the Leyden jar (a foil-lined glass jar), a single, isolated metallic plate, and a multiple (fixed or variable) plate condenser.

Electrostatic induction occurs when a charged object is brought near a neutral grounded object. If the ground connection is broken before the charged object is removed, the charge driven off through the ground wire cannot return, and the previously neutral object is left with a "permanent" induced charge. An *electroscope* is an instrument which indicates a charge without measuring it.

8 ELECTRIC CURRENT

SELF-TEST

1. Several chemical cells producing electricity may be combined to form a(an) _____.

2. The chemical cell produces _____ current.

3. Current is usually produced by chemical generation or electromagnetic _____.

4. Relative motion between a conductor and a magnetic field induces current. This is the principle of the _____.

5. Generation of electricity by light is accomplished by the _____.

6. Generation of electricity by heat is accomplished by the _____.

7. A current through a conductor in a magnetic field induces motion. An application of this effect is the electric _____.

8. The opposite of chemical generation of a current by a cell or battery is chemical _____.

9. A generator basically produces a(an) _____ current.

10. Electric current will flow only if the circuit path is _____.

11. In Ohm's Law, current = electromotive force ÷ _____.

12. The unit of electric current is the _____.

13. The unit of electric "pressure" is the _____.

14. The unit of electric resistance is the _____.

15. If the right hand clasps a conductor wire so that the fingers point in the direction of the lines of force, the thumb points in the direction of the _____.

16. Current flow is conventionally drawn towards the _____ pole.

17. A fuse is inserted into a circuit so that, if overheating occurs, it will be the first part of the circuit to _____.

18. Two close, but not connecting, coils that transfer an electric current by induction form a(an) _____.

1 _____
2 _____
3 _____
4 _____
5 _____
6 _____
7 _____
8 _____
9 _____
10 _____
11 _____
12 _____
13 _____
14 _____
15 _____
16 _____
17 _____
18 _____

29

BASIC FACTS

An electric current is a flow of electrons (or ions) through a conducting medium.

MAJOR METHODS OF GENERATION OF ELECTRIC CURRENT. *Chemical generation* is the production of an electric current by a chemical effect.

The basic unit for the chemical generation of an electric current is the *chemical cell*, of which there are three important types: the primary wet cell (in which the electrolyte is a solution), the primary dry cell (in which the electrolyte is a paste), and the rechargable (storage) cell.

Several cells are combined to form a *battery*.

Electromagnetic induction, the principle of the generator, is the production of an electric current by the relative motion of a conductor and a magnetic field.

Solenoid coils used with a bar magnet and the *AC generator* are examples of two devices that generate an electric current by electromagnetic induction.

In the generator, mechanical energy is converted to electrical energy.

MINOR METHODS OF GENERATION. The *photoelectric effect* is the emission of electrons from a metal plate struck by light of a sufficiently high frequency. This is the principle of a solar battery. In a *photocell*, the electrons thus emitted travel to a positively charged plate, and an electric current is generated.

A *thermocouple* is two wires of different metals joined at both ends. If such a metal pair is heated at one

(Continued on page 32)

ADDITIONAL INFORMATION

A *chemical cell* consists of two terminals in an ionized solution (electrolyte). Solution ions remove ions of opposite charge from the originally neutral terminals. This leaves the terminal with the same charge sign as its solution ion had. A SO_4^{-2} ion in solution, for example, will combine with a Zn^{+2} or Pb^{+2} ion from a terminal, leaving it with a negative charge. If a wire connection is made between the terminals outside the cell, electrons will flow through the wire to the positive terminal, thus completing the circuit. This produces a steady *direct current* (DC).

In an *electrolytic cell*, electric current is used to produce a chemical effect, such as the electroplating of metals.

Exchange of electrons occurs at the *poles* in both chemical and electrolytic cells. The *anode* is the pole where: oxidation (loss of electrons) occurs, and electrons are removed from negative ions in the solution. The *cathode* is the pole where: reduction (a gain of electrons) occurs, and electrons "enter" the solution. The anode is positive in an electrolytic cell and negative in a battery; the reverse is true for the cathode. In an electrolytic cell, negative ions (anions) are attracted to the positive pole, and positive ions (cations) to the negative pole. In a chemical cell, negative ions move toward the negative pole, and positive ions toward the positive pole, thus inducing a flow of current.

Electromagnetic induction occurs when a wire conductor is coiled into several loops (*solenoid coils*) and a bar magnet is inserted into the coils. The relative motion of both the coils and the magnet (or either one) induces a current in the coils.

The *AC generator* consists of a magnet between whose poles a coil of wire is rotated. During half the rotation, the magnetic lines of force are cut in one direction, and a current is produced in the coil. During the other half of the rotation, the lines of force are cut in the opposite direction, and the direction of the current is reversed. The current thus produced is an *alternating current* (AC). If no connection with an outside circuit exists during half the wire's rotation, a *periodic direct current* is produced. Commutator rings, which reverse connections to the coil every half-rotation, allow a pulsating direct current. In commercial generators, the mechanical energy is usually provided by steam or water power.

In the *motor*, two *electromagnets* are mounted so that one of them (the armature) is free to turn in the magnetic field of

(Continued on page 32)

EXPLANATIONS

1. Most car batteries have now been increased from 6 to 12 volts, primarily by increasing the number of cells.

2. A chemical cell produces a direct (flowing in one direction) and steady current. Alternating current converted to DC is direct but periodic, unless "smoothed out" by special circuitry.

3. The term induction implies that something is not taking place by direct contact. There is no connection between the magnet and the conductor other than the intangible lines of force.

4. The principle of the generator involves relative motion. It does not matter whether one part or the other (or both) move.

5. The photocell or "electric eye" contains a substance which generates a small, steady, direct current when light falls on it.

6. The thermocouple (composed of two dissimilar metals) generates a small, steady, direct current when its junction is heated.

7. A current that induces motion forms a motor effect.

8. In electrolysis a current produces a chemical effect; in a battery a chemical effect produces current.

9. A generator (unless converted to DC by commutator rings) produces AC or *alternating* current. This flows first in one direction and then the other (in the case of household current, usually reversing about 60 times a second).

10. An unbroken as opposed to an open circuit.

11. Ohm's law states that the current equals the electromotive force divided by the resistance (in a given circuit).

12. The ampere is named after André Marie Ampère.

13. The volt is named after Alessandro Volta.

14. The ohm is named after G. S. Ohm.

15. The right hand rule of the conductor wire may be used to indicate *either* the direction of lines of force *or* the direction of current flow.

16. In reality, current usually flows in the opposite direction (negative electrons towards the positive pole). The convention of positive charges moving toward a negative pole predates the discovery of the electron. (In some instances, this may actually be the case—for example, ions in solution.)

17. A fuse thus prevents fires and protects the more costly circuit components.

18. A transformer may be used to "step up" or "step down" the voltage as desired, depending on the ratio of the number of turns in the two coils.

Answers

battery	1
direct	2
induction	3
generator	4
photocell	5
thermocouple	6
motor	7
electrolysis	8
alternating	9
unbroken	10
resistance	11
ampere	12
volt	13
ohm	14
current	15
negative	16
melt	17
transformer	18

Ohm's Law Mnemonic

E	
I	R

Use of the Mnemonic: To find top half, multiply two bottom quarters. To find one bottom quarter, divide top half by other bottom quarter.

junction, and kept cold at the other, an electric current will flow between the junctions. Because the current strength is directly related to temperature difference, thermocouples are used in the measurement of temperature.

ELECTRIC CIRCUITS. An *electric current* will flow only if there is an unbroken path (*circuit*) from, and returning to, the source of the current.

A circuit's *electromotive force* (*E*) is the voltage across its terminals when no current is being drawn.

A circuit's *potential difference* (*V*) is the voltage drop when a current is being drawn.

Current (*I*) is the flow of electrons along a conductor.

Resistance (*R*) is the internal opposition to the passage of an electrical current.

Ohm's Law:

$$\text{current} = \frac{\text{electromotive force}}{\text{resistance}}$$

A *series circuit* has its parts so connected that the entire current passes through each part, in turn.

A *parallel circuit* has its parts so connected that the current is divided between each of the parts.

STANDARD SYMBOLS. In *circuit diagrams*, the direction of current flow is *conventionally* drawn from the positive terminal to the negative terminal, even though it is known that an electron will not flow to a negative terminal.

The direction relationships of electromagnetics can be determined by the right hand rule of the conductor wire, the right hand rule of the generator, and the left hand rule of the motor.

A *transformer* is used to raise or lower the voltage in AC circuits.

POWER AND ENERGY. *Electric power* (watts) = $I \times E$ *Energy* (watt-hours) = $t(IE)$, where t stands for the elapsed time.

the other (the yoke). When current flows, the electromagnets exert magnetic force, and the armature turns to match its poles to those of the yoke. If it did this, though, it would stop. Just before the poles match, the polarity of the yoke is changed, and the armature makes another half turn. This process repeats itself as long as the current flows.

In *Ohm's Law*, electromotive force is measured in volts, resistance (*R*) is measured in ohms, and current (*I*) is measured in amperes. Ohm's Law may be used to solve for any one of *E*, *I*, or *R* by rewriting the formula according to elementary algebra:

$$I = \frac{E}{R} \qquad E = IR \qquad R = \frac{E}{I}$$

Since all of the current in a *series circuit* passes through each resistor, the total resistance in the circuit is the sum of the individual resistances: $R = R_1 + R_2 + R_3$.

In a *parallel circuit*, the total resistance is the sum of the reciprocals of the individual resistances: $R = \frac{1}{R_1} + \frac{1}{R_2} + \frac{1}{R_3}$, etc.

The right hand rule of the conductor wire—If the hand clasps around the wire with the thumb in the direction of the current, the fingers circle the wire in the direction of the magnetic field.

The *right and left hand rules of the generator and motor*—Extend the thumb, first finger, and middle finger all at right angles to each other, the first finger pointing ahead. The thumb shows direction of motion, the first finger shows direction of field and the middle finger shows direction of current. The right hand describes the generator, and the left hand describes the motor.

The *galvanometer* indicates the presence or absence of a current. The *ammeter* measures, in amperes, the amount of current flow. It is a low resistance shunt wire connected in parallel with a measuring galvanometer. Most of the current goes through the shunt and the ammeter measures the fractional portion that does not. The *voltmeter* measures, in volts, the electrical pressure behind the current. It is a high resistance in series with a measuring galvanometer, the whole section being connected into the circuit in parallel. The *combined voltammeter* measures both voltage and current. The *Wheatstone bridge* measures an unknown resistance by balancing it against three known resistances. The *potentiometer* does the same thing, using batteries instead of resistance wires.

Fuses contain a low resistance wire which will melt before any other part of the circuit. *Circuit breakers* trip a switch to open a circuit.

9 KINDS AND PROPERTIES OF SUBSTANCES

SELF-TEST

1. Properties that are due to (or dependent on) the kind of matter present are
   a. extrinsic
   b. intrinsic
   c. altruistic

2. A change where no new substances are formed is a
   a. physical change
   b. chemical change

3. A reaction where heat is given off is called
   a. endothermic
   b. exothermic

4. The principal atomic particle found in the nucleus and possessing a positive charge is called the
   a. electron
   b. neutron
   c. proton

5. The atomic particle that has a negative charge and orbits around the nucleus is called the
   a. electron
   b. neutron
   c. proton

6. The number of protons in the nucleus of an atom is the atom's
   a. atomic weight
   b. atomic number
   c. shell structure

7. The number of protons and neutrons in the nucleus of an atom is the atom's
   a. atomic weight
   b. atomic number
   c. shell structure

8. A molecule with two atoms is
   a. monotomic
   b. diatomic
   c. polytomic

9. Two or more different atoms bonded together defines a
   a. molecule
   b. compound

10. H_2, N_2, O_2, and F_2 are
    a. molecules
    b. compounds
    c. both

11. H_2O, H_2SO_4, and HNO_3 are
    a. molecules
    b. compounds
    c. both

12. A non-homogenous combination of two or more substances in indefinite proportions is a
    a. compound
    b. mixture
    c. solution

13. Small numbers that tell how many atoms of each element are in a compound are called
    a. subscripts
    b. valences
    c. compound numbers

1 _____
2 _____
3 _____
4 _____
5 _____
6 _____
7 _____
8 _____
9 _____
10 _____
11 _____
12 _____
13 _____

BASIC FACTS

KINDS OF PROPERTIES. The *extrinsic properties* of a substance are independent of the material of which the substance is composed. The *intrinsic properties* are due to the material of which the substance is composed.

Intrinsic properties are either physical or chemical properties: observation does not change physical properties, it can change chemical properties.

KINDS OF CHANGES. In a *physical change*, no new substances are formed. In a *chemical change*, at least one new substance is formed.

An *endothermic reaction* is one which can occur only if heat is added. An *exothermic reaction* is one in which heat is given off.

An *atom* is the smallest particle into which a substance can be divided chemically (without change in the nucleus).

PARTICLES OF THE ATOM. The most important particles of the atom are the *proton*, the *neutron*, and the *electron*. The atom consists of a solid *nucleus* with most of the weight (protons and neutrons) at the center. Electrons rapidly circle the nucleus at various levels or shells located at comparatively great distances from the nucleus.

PROPERTIES OF THE ATOM. The atomic number is the number of protons in the nucleus of an atom. The atomic weight is the total number of heavy particles (protons plus neutrons) in the nucleus. The electron shells are indicated (according to their distance from the nucleus) by the letters K, L, M, N, O, and P. The number of electrons for each shell are: $K(2)$, $L(8)$, $M(18)$, $N(32)$, and, to date, $O(18+)$ and $P(8+)$.

An *element* is a substance all of whose atoms have the same atomic number. To date, 103 elements have

(Continued on page 36)

ADDITIONAL INFORMATION

Chemistry is the science which considers the composition of matter and the changes in its composition. The *extrinsic properties* of matter include shape, length, weight, and temperature. Some *physical properties* are odor, density, hardness, and freezing and boiling points. Some *chemical properties* are reactions to acids and bases, and degree of ionization.

Some typical *exothermic* reactions are combustion, slow oxidation (such as rusting and decay), and spontaneous combustion (combustion without outside ignition).

Basic Particles of the Atom

Name	Symbol	Location	Relative Mass	Charge
Electron	e^-	in orbit	$\frac{1}{1836}$	−
Proton	p^+	in nucleus	1.0072	+
Neutron	n^0	in nucleus	1.0087	0

Note that for both mass and charge, electron + proton = neutron.

In a neutral atom, the number of electrons (−) must be equal to the number of protons (+). The speed of an electron's motion tends to hurl it outward along a tangent to the elliptical path in which it travels; but the attraction between the positive nucleus and the negative electron counterbalances the electron's tendency to fly off and keeps it in orbit. The paths of the electrons outline the surface of three-dimensional figures (for example, a sphere instead of a circle).

The *atomic number* is a convenient orderly method of distinguishing all the atoms of different elements by their different number of protons.

No *outermost electron shell* ever has more than eight electrons. The innermost shell, the K shell, never has more than two electrons. The electrons in the outer shells are very important because most of the properties of the atom depend upon them (except weight—almost all of the atom's weight is in the nucleus). The only parts of the atom that enter into chemical reactions are these outer electrons. The compounds formed by atoms, and the atoms' ability to ionize, are governed by the number of outer-shell electrons. Electron shells between the innermost and outer shells may have more than eight electrons.

The Element Concept has changed. Ideas about elements have undergone important changes as man has come to learn more about his physical environment. Such apparent things as earth, water, air, and fire were first considered basic and called elements. For almost 2000 years, men sought to

(Continued on page 36)

EXPLANATIONS

1. *In-* means into or concerned with; *ex-* means out of or away from. Extrinsic properties are independent of the material of which the substance is composed.

2. An example of a physical change is the change from ice to water to steam. The formula of the individual molecules in all three states is H_2O.

3. Analogously, a reaction to which heat must be added is an endothermic reaction.

4. The mass of a proton is approximately equal to the mass of neutron (which has a neutral charge).

5. The mass of an electron is approximately 1/1836th that of a proton.

6. Each element has only one possible atomic number.

7. An element can have more than one atomic weight, depending on the number of neutrons present in the nucleus.

8. Similarly, a molecule consisting of one atom is monatomic, and a molecule containing more than two atoms is polyatomic.

9. In order to be a compound, the molecule must have two *different* atoms.

10. All the substances listed have two similar atoms. Hence, they are not compounds.

11. All the substances listed are compounds as they have two or more dissimilar atoms. Because all compounds are molecules (although not all molecules are compounds), these substances are also molecules.

12. A mixture is not homogeneous down to molecular size, while a compound is.

13. Subscripts are the small whole numbers printed at the lower right side of the symbol for an element.

Answers	
intrinsic	1
physical change	2
exothermic	3
proton	4
electron	5
atomic number	6
atomic weight	7
diatomic	8
compound	9
molecules	10
both compounds and molecules	11
mixture	12
subscripts	13

been discovered or prepared; 92 of which occur in nature. The rest have been made in the laboratory since 1940.

MOST ABUNDANT ELEMENTS. Hydrogen is the most abundant element in the universe. Nitrogen gas accounts for 78% of the atmosphere (by volume). Oxygen (49%) and silicon (26%) make up 75% of the earth's crust (by weight). Oxygen (65%), carbon (18%), and hydrogen (10%) make up over 90% of living matter.

DESIGNATION OF ELEMENTS. *Element Symbols* are usually composed of the first one or two letters of the English or Latin names of the elements: for example, H for hydrogen, He for helium, C for carbon, Fe for iron.

Formulas are combinations of symbols used to represent the composition of molecules or compounds: for example, H_2, F_2, H_2O, Fe_2O_3.

Subscripts are small numbers written at the lower right hand corner of a symbol. In chemical formulas they indicate the number of atoms of that element in the molecule. By convention, atoms with a positive valence are generally written first in a formula.

MOLECULES AND COMPOUNDS. A *molecule* is two or more atoms bonded together. A *monatomic* molecule contains only one atom, a *diatomic* molecule contains two atoms, and a *polyatomic* molecule contains more than two atoms.

A *compound* is two or more *different atoms* bonded together. Over a million compounds are known.

THE LAW OF DEFINITE PROPORTIONS (LAW OF CONSTANT COMPOSITION). Every sample of a compound contains the same elements in the same proportions by weight.

change one element to another by adding or subtracting properties such as hotness, coldness, wetness, and dryness. Others thought that matter must be comprised of extremely small basic units that were all the same, but combined in different numbers and arrangements to form different substances. This view was revived in the last two centuries. However, it is the subatomic particles (protons, neutrons, and electrons) that are the basic building blocks, rather than what are now called elements.

If a symbol in a *formula* has no *subscript*, the subscript 1 is implied. An implied 1 is never written. For example, write $NaCl$—not Na_1Cl_1.

Valence refers to the combining capacity of an atom or radical (group of atoms functioning as a unit in a chemical change), using hydrogen with a valence of $+1$ as a standard. (Oxidation number, the apparent charge on an atom in a compound, is sometimes used synonymously with valence.) Subscripts and valences are related. In compounds of two elements, valence numbers (without signs) may sometimes be crossed to become subscript numbers and vice versa, allowing prediction of one from the other. For example:

$$Fe^{+3} \diagdown\!\!\!\!\diagup O^{-2} \qquad +3 \diagdown\!\!\!\!\diagup -2 \qquad \text{Valence Numbers}$$
$$Fe_2O_3 \qquad\qquad 2 \diagdown\!\!\!\!\diagup 3 \qquad \text{Subscript Numbers}$$

A *molecule* is sometimes given an alternate definition and described as the smallest particle of a compound that can exist in nature and still retain the properties of that compound. Sodium (Na) reacts violently with water, and chlorine (Cl) is a poisonous gas. Still, we are able to eat table salt (NaCl) because its properties differ from those of either sodium or chlorine.

DISTINCTION BETWEEN MOLECULES AND COMPOUNDS. All compounds are molecules, but not all molecules are compounds. For example, $NaCl$, H_2O, and H_2SO_4 are both molecules and compounds, but H_2, O_2, N_2 are not compounds, since their atoms are not different.

DIFFERENCES BETWEEN MIXTURES AND COMPOUNDS.
(1) A compound is of the same composition throughout, while the composition of a mixture varies. We can add more of any one constituent of a mixture.
(2) The proportions of each substance in a compound are fixed. They are not fixed in a mixture.
(3) A compound has only one set of characteristic properties, while a mixture has no common set of characteristic properties.

10 PRACTICAL UNITS AND OXIDATION STATES

SELF-TEST

1. The _total_ of all the atomic weights in a compound is the molecular weight of the compound.

2. If the molecular weight of O_2 is 32, its gram molecular weight is _32 gms_.

3. The atomic weight of hydrogen is 1, and the atomic weight of oxygen is 16. The molecular weight of H_2O is _18_.

4. The number of electrons lost, gained, or shared by an atom is the _valence_ of the atom.

5. If the number of outer shell electrons of an atom is under 4, the sign of its valence is _+_.

6. Sulfur has six electrons in its outer shell. Its valence is _-2_.

7. Neon (an inert gas) has a valence of _0_.

8. In reduction, an atom _gains_ electrons.

9. Atoms of the same element that differ only in their atomic weights are called _isotopes_.

10. In an oxidation-reduction reaction, the reducing agent is the element that is _oxidized_.

11. Salt dissolved in water, "carbonated" water, and alloys are examples of _____.

12. The oxidation state of oxygen is almost always _____.

13. The type of valence where one atom donates all the shared electrons is _____.

14. If a current will flow through the water solution of a substance, the type of bond found in the substance is _____.

15. The total valence of a compound is _____.

BASIC FACTS

MIXTURES AND SOLUTIONS. A *mixture* is a combination of two or more substances in indefinite proportions; it is not homogeneous down to molecular size. A *solution* (homogeneous mixture) is a special type of mixture. Its proportions are not fixed, but are the same throughout. It exhibits uniformity down to microscopic, but not molecular, size. The *solvent* is the substance which dissolves another substance. The *solute* is the substance which is dissolved.

ATOMIC VARIATIONS. An *ion* is an atom which has gained or lost electrons. It therefore bears a charge. Atoms of the same element having different atomic weights (different numbers of neutrons) are called *isotopes*. An isotope has "gained" or "lost" neutrons.

A *radical* is a complex ion; a group of elements possessing a common charge and moving in an equation as a unit. (For example, the positive ion NH_4^+ or the negative ions OH^-.) In a radical, at least one atom must be positive and one atom must be negative.

MOLECULAR WEIGHTS. The molecular weight of a molecule is the sum of *all* its atomic weights. The molecular weight of a compound is calculated by taking the sum of the products of the atomic weights of each element present in the compound multiplied by the subscript of that element.

THE PRACTICAL UNIT SCALE. The *gram atomic weight* (G.A.W.) of an element is the weight of that element in grams which is numerically equal to the element's atomic weight. Similarly, the *gram molecular weight* (G.M.W.) of a compound (molecule)

(Continued on page 40)

ADDITIONAL INFORMATION

The most common *solutions* are solid-in-liquid solutions (for example, sugar in coffee or salt in water) and gas-in-liquid solutions (for example, CO_2 gas in a "carbonated" soft drink). There are also liquid-in-liquid solutions (for example, alcohol in water) and solid-in-solid solutions (for example, alloys or solder).

The *solvent* is generally the substance present in the greatest amount. For example, in a solution of 60% alcohol and 40% water, alcohol is the solvent. An alloy is formed by pouring two metals together while hot and allowing them to cool.

An *ion* has a different number of electrons than protons. Many substances will dissociate in solution to form ions. These substances are said to ionize, and are called *electrolytes*. The most common solvent is water.

There are two types of *ionic cells:* electrolytic cells and chemical cells (or batteries). The poles (anode and cathode) are places of exchange of electrons in ionic cells (See Chapter 8).

The *atomic weight* is a pure number scale (without units) relative to some standard. One element is chosen as the standard, and the weights of other elements are determined in comparison to it. Hydrogen (AW = 1.0), oxygen (AW = 16.0), and carbon (AW = 12.0) have been successively used as the standard. The atomic weight of a pure isotope would have to be a whole number because only whole protons or neutrons can be added or subtracted from the nucleus. The atomic weights listed in most tables, however, are decimals because they represent the statistical average of an actual sample and any sample would be a mixture of the various isotopes of an element.

When calculating *molecular weights*, the atomic weights of all atoms present must be included. For example:

M.W. of Fe_2O_3 = 2 (A.W. of Fe) + 3 (A.W. of O).
M.W. of H_2O = 2 (A.W. of H) + 1 (A.W. of O).

Gram atomic weight has the chemical convenience that, if we take the number of grams of each of two substances that correspond to their atomic weights, we have an equal number of molecules in each sample. For example, the atomic weight of hydrogen is 1.008 and the atomic weight of oxygen is 15.999. Thus, the gram atomic weight of hydrogen is 1.008 g and the gram atomic weight of oxygen is 15.999 g. Therefore, 1.008 g of hydrogen has the same number of molecules as 15.999 g of oxygen. This is helpful in figuring

(Continued on page 40)

Practical Units and Oxidation States 39

EXPLANATIONS

1. The simplest method of calculating the molecular weight of a compound is to multiply the atomic weight of each element present by the subscript of that element, then add the products.

2. The gram molecular weight of oxygen gas is the number of grams of oxygen that is numerically equal to the molecular weight of the oxygen molecule.

3. M.W. of H_2O = 2(A.W. of H) + (A.W. of O) = 2(1) + 16 = 18

4. The valence of an atom can also be defined as the combining capacity of an atom relative to hydrogen (with a valence of +1).

5. The atom most readily loses these electrons, thus giving it a positive valence.

6. As the atom has six electrons (more than four), it will gain two more electrons, giving it a valence of −2.

7. All the inert gases have a full outer electron shell.

8. Remember the mnemonic of growling: grr—gain equals reduction.

9. Different isotopes of an element have different numbers of neutrons.

10. The reducing agent is so named because it reduces the other reactant. Likewise, the oxidizing agent is the element that is reduced.

11. These are, respectively, a solid-in-liquid solution, a gas-in-liquid solution, and a solid-in-solid solution.

12. The principal exception to this rule is the peroxides (such as H_2O_2), where the valence of oxygen is −1.

13. Covalent bonding involves a sharing of electrons between atoms.

14. In ionic bonding, individual ions become free to move toward the oppositely-charged pole, thus passing a current.

15. Because a compound is electrically neutral, the sum of the positive and negative charges must be zero.

Answers	
sum	1
32 g	2
18	3
valence	4
positive	5
−2	6
0	7
gains	8
isotopes	9
oxidized	10
solutions	11
−2	12
coordinate covalence	13
ionic	14
0	15

is the number of grams of that compound which are numerically equal to its molecular weight. The weight of one mole (Avogadro's number, 6.02×10^{23}) of atoms (molecules) is equal to the G.A.W. (G.M.W.) of a given atom (molecule).

ELECTRON ACCOUNTING. Two methods are used: valence and oxidation states.

The *valence* of an atom is:

1. The combining capacity of an atom or radical compared to hydrogen (with a valence of $+1$).
2. The number of electrons lost, gained, or shared.

Valences are usually positive if the number of electrons in the outer (valence) shell is under four, and negative if over four. The valence number is usually the number of electrons in the outer shell (if there are less than four of these), or eight minus the number of electrons in the outer shell (if there are over four).

OXIDATION STATES. *Oxidation* occurs when an atom loses electrons. *Reduction* occurs when an atom gains electrons. In order for an atom to be oxidized, some other atom must be reduced. The apparent charge on an atom is its *oxidation number* (oxidation state). Oxidation numbers range from 0 to $+7$.

Both valence and oxidation number are determined by the number of electrons in the outer shell of the atom. Both terms are often used interchangeably.

The valence, or oxidation number, of a free and uncombined element is zero. The sum of all the oxidation numbers of the elements in a compound is equal to zero.

combining proportions and in converting from weight to volume.

When calculating *gram molecular weight*, all the gram atomic weights present must be considered. For example, G.M.W. of $H_2O = 2(1.008 \text{ g}) + 15.599 \text{ g} = 18.015 \text{ g}$.

In the calculation of valence, it should be remembered that atoms tend to gain or lose electrons in such a manner as to attain a full outer electron shell of eight electrons. If there are four, or less than four, electrons in the outer shell of an atom, the atom is likely to lose these electrons—giving it a positive valence. For example, an atom with two electrons in its valence shell will lose these electrons to attain a valence of $+2$. If there are more than four electrons in the atom's outer shell, the atom is likely to gain electrons—giving it a negative valence. For example, an atom with seven electrons in its outer shell would tend to gain one additional electron to attain a valence of -1. Atoms which already have a full outer shell of eight electrons (the inert, noble or rare, gases) show almost no tendency to form compounds.

In an oxidation-reduction reaction, the reactant which is oxidized (loses electrons) is called the reducing agent (because it reduces the other reactant). The reactant which is reduced (gains electrons) is called the oxidizing agent (because it oxidizes the other reactant). For every atom that loses electrons in an oxidation-reduction reaction, some atom(s) must gain an equal number of electrons.

Elements may exist in *alternate oxidation states*. That is, an element may have one oxidation state in one compound but a different oxidation state in another compound. Two elements whose oxidation states are almost always the same are hydrogen ($+1$) and oxygen (-2). Exceptions are metal hydrides ($H = -1$) and peroxides ($O = -1$).

The *valence of a radical* is the algebraic sum of the oxidation states of its elements.

TYPES OF VALENCE. *Ionic valence* involves a transfer of electrons from one atom to another. One atom actually gives up electrons, while another atom receives them. *Covalence* involves a sharing of electrons. No actual physical separation of an electron from its atom occurs. In coordinate covalence, one atom donates all the shared electrons. In polar covalence, two atoms of different elements (nonidentical atoms) both donate shared electrons. In nonpolar covalence, two atoms of the same element (identical atoms) both donate shared electrons.

11 THE PERIODIC TABLE

SELF-TEST

1. The modern Periodic Table lists all the elements in order of their increasing atomic _number_.

2. Mendeleev's original Periodic Table listed all the elements he knew of in order of their increasing atomic _wt_.

3. The vertical columns of the Periodic Table are called _group_.

4. The number of electrons in the outer shell of every one of the elements in each of these vertical columns is _the same_.

5. The horizontal rows of the Periodic Table are called _periods_.

6. The greatest tribute to Mendeleev was the gradual accurate fulfillment of his _____ _____.

7. The B groups of the modern Periodic Table are called the _transition_ elements.

8. In the B groups, electrons are added to _inner_ shells.

9. The class of substances that is characterized by a readiness to give up electrons is called _metals_.

10. An element will displace from solution any element placed _below_ it in the activity series table.

11. Elements in Groups III A, IV A, and V A are called _metaloids_.

12. The elements on the left hand side of the Periodic Table are collectively termed _metals_.

13. The Group VII A elements are called the _halogens_.

14. The Group I A elements are called the _alkali_ metals.

15. The modern version of the Periodic Table, which is spread out, is called a _____ table.

16. An element in Group VI A will have _6_ electrons in its outer shell.

17. An element in Group VI A may assume a valence of _−2_.

41

BASIC FACTS

The *Periodic Table* of elements forms the basis for modern chemistry. It stems from the work of Mendeleev.

Mendeleev's Periodic Law states that when the elements are arranged in order of increasing atomic weight, their properties are repeated periodically. From this he created the first Periodic Table, in which elements with similar properties were arranged in vertical columns.

This repetition led to Mendeleev's predictions of the existence of undiscovered elements, and of their properties, which have been fulfilled to a high degree of accuracy.

Modern Periodic Tables use atomic number instead of atomic weight.

MAJOR DIVISIONS OF THE TABLE. The *groups* are vertical columns, representing elements which have similar properties and valences. The elements in each column have the same number of electrons in their outer shells.

The *periods* are horizontal rows of increasing atomic number. Atomic weight (with few exceptions) and the number of electrons also increase from left to right.

For better correlation of properties, the modern long form table has been spread out and split into *A* and *B* sections. The *B* Section (the transition elements) is comprised of the right hand subgroups from the old short form table. They are placed between Groups II *A* and III *A* in the modern table.

Roughly vertical lines can be drawn down the chart separating light metals, heavy metals, and nonmetals.

(Continued on page 44)

ADDITIONAL INFORMATION

In 1869, Dimitri I. Mendeleev, a professor of chemistry in Russia, proposed a *periodic classification* of the 63 elements then known to exist, and formulated his *Periodic Law*.

The *Periodic Table* is a device for the classification of elements. Modern periodic tables use the atomic number (number of protons), rather than the atomic weight, as the basis for classifying the elements. In Mendeleev's day, protons had not been discovered.

In constructing his table, Mendeleev had to leave blank spaces. He assumed that these spaces represented elements which had not yet been discovered. By comparison with the properties of known elements in the vertical rows of his table, Mendeleev was able to predict the chemical and physical properties of the undiscovered elements. It took over fifty years to fill in the blank spaces in Mendeleev's table.

Even within his vertical rows (Groups), he recognized differences, so he made right- and left-hand subgroups.

The groups indicate valences, because the group number (in the *A* Section) is also the number of electrons in the outer shell.

Group	I	II	III	IV	V	VI	VII	VIII
Primary Valence	+1	+2	+3	±4	−3	−2	−1	0

Group No. <4, Valence = Group No.
Group No. >4, Valence = 8−Group No.

The groups increase in number of electron shells from top to bottom.

Family names have been given to some commonly used chemical groups. Group I *A* (includes lithium, sodium, and potassium) is called the alkali metals. Group VII *A* (includes fluorine, chlorine, bromine, and iodine) is called the halogens.

Metals are usually solids; compact, dense, and good conductors of heat and electricity. They do not readily combine with each other, but they combine with nonmetals readily. Their oxides form bases. Hence, the metal oxides are called basic anhydrides.

Nonmetals are usually solids or gases (a few are liquids). They are light weight and are poor conductors. Some com-

(Continued on page 44)

EXPLANATIONS

1. This is because the number of protons is the same for every isotope of an element.

2. Mendeleev listed atomic weights, as protons were not known in his time.

3. The vertical columns of the table are called groups by definition.

4. Because the elements in each group have the same number of valence electrons, they have similar properties.

5. The horizontal columns of the table are called periods by definition.

6. The work of many men, over a period of more than 50 years, was required to verify Mendeleev's predictions.

7. The B groups, altogether, make up the section.

8. In the B groups, electrons are usually added to the next-to-outermost shell, while the outer shell continues to hold two electrons.

9. A tendency to give up electrons is characteristic of metals.

10. In the activity series table, elements displace other elements found below them and are displaced by the ones above them.

11. The "borderline elements," Groups III A, IV A, and V A are called metalloids by definition.

12. Metals are found on the left side of table, nonmetals are found on the right side of table.

13. The halogens (salt formers) include fluorine, chlorine, bromine, and iodine.

14. The alkali metals include sodium and potassium. These very light, active metals form the common salts which accumulate on alkali salt flats.

15. The modern Periodic Table is spread out and split in order to place all the transition elements together in the B section.

16. For elements in the A section, the number of electrons in the outer shell of an atom is the same as the group number of the atom.

17. If the number of electrons in the outer shell of an atom is greater than four, then the characteristic valence for the atom is eight minus the group number, and the valence is negative.

Answers

numbers	1
weights	2
groups	3
the same	4
periods	5
predictions	6
transition	7
inner	8
metals	9
below	10
metalloids	11
metals	12
halogens	13
alkali	14
long-form	15
6	16
− 2	17

A periodic table. Roman numerals designate groups; arabic, periods. Atomic numbers are listed above and left of element symbols; atomic weights, below.

METALS are substances that are generally hard, lustrous, and conductive and will give up electrons. They are located mostly on the left side of the Periodic Table.

NONMETALS are nonconductive substances that lack luster and will take on electrons. They are located on the right side on the Periodic Table.

METALLOIDS. The metalloids are the borderline elements. They are found in Groups III A, IV A, and V A of the Periodic Table. They may act as metals or nonmetals.

ELECTRON SHELL STRUCTURE. The A and B Sections are arranged in order of addition of orbital electrons. The A Section elements add electrons to outer shells. The B Section elements add electrons to an inner shell, while the number of electrons in the outer shell generally remains at two. The *lanthanide series* (rare earth) and the *actinide series* fit into the table only as footnotes. Lower-level *subshells* are being filled in both series. The chemically inactive inert gases (also called rare gases or noble gases) have eight outer electrons and are placed in a column at the right of the table. They were unknown and unpredicted by Mendeleev.

bine readily with each other, but most combine more readily with metals. Their oxides form acids. Hence, the nonmetal oxides are called acid anhydrides.

ACTIVITY SERIES OF ELEMENTS. Any element in an activity series is more active than, and will displace from solution, any element listed below it. Two series exist: (1) metals and hydrogen and (2) nonmetals.

In general, the metals tend to be more active towards the bottom of a group. In the larger atoms of elements near the bottom of a group, the outer electrons are farther from the nucleus and less tightly held. Nonmetals tend to be more active towards the top, where the outer electron shell will be closer in and additional electrons will be more strongly attracted.

An outer shell never contains more than eight electrons, as long as it is an outer shell. The innermost shell (K) never contains more than two electrons, being too close to the nucleus to hold more.

The second (or L) shell starts with lithium (Li), whose third electron starts the new shell (2-1 configuration). This shell becomes filled with the inert gas neon (Ne), which has a 2-8 configuration.

The third (or M) shell starts with sodium (Na), which has a 2-8-1 configuration, and continues through the third period to the inert gas argon (Ar), 2-8-8.

The fourth (or N) shell starts with potassium (K), 2-8-8-1. The next element calcium (Ca) is 2-8-8-2. At this point something different happens. The third element in the fourth shell should be 2-8-8-3, but it is not. It is 2-8-9-2! Since the third shell is now an inner shell, it continues filling up to zinc (Zn), 2-8-18-2. The fourth period (or horizontal row) continues through to the inert gas krypton (Kr), 2-8-18-8. The N shell is not finally completed until the rare earth ytterbium (Yb, atomic number 70), which is 2-8-18-32-8-2.

Each shell may be divided into s, p, d, and f subshells on the basis shape (and, hence, energy). The maximum number of electrons permissable in each subshell is: s (2), p (6), d (10), and f (14). Of the eight possible M electrons, two must be s electrons ($3s$), and six must be p electrons ($3p$).

12 CHEMICAL CLASSIFICATION

SELF-TEST

1. For laboratory safety always add
    a. acid to water
    b. water to acid
    c. both at the same time

2. If a solution feels slimey and turns red litmus blue it is most likely
    a. an acid
    b. a base
    c. a salt
    d. water

3. If a solution tastes sour and turns blue lutmus red it is most likely
    a. an acid
    b. a base
    c. a salt
    d. water

4. If a solution has a $pH > 7$, it is most likely
    a. an acid   less than
    b. a base
    c. a salt
    d. water

5. A neutralization reaction produces
    a. a salt
    b. water
    c. a base
    d. both a and b

6. If the pH of a solution is 14, the solution is
    a. a strong acid
    b. a strong base
    c. a salt
    d. water

7. The type formula for a series of compounds is $C_n H_{2n+2}$. How many hydrogens does the member of this series with two carbons have?
    a. 6
    b. 8
    c. 24

8. The higher acid "valence" ending is
    a. *-ous*
    b. *-ic*
    c. *-ite*

9. A prefix for a binary (two element) acid is
    a. *hydro-*
    b. *hypo-*
    c. *per-*

10. A lower-valence, ternary (3 element) salt ending is
    a. *-ide*
    b. *-ite*
    c. *-ate*

11. The compound Fe_2O_3 is named
    a. ferrous oxide
    b. ferric oxide
    c. plumbic oxide

12. $C_n H_{2n}$ is an example of a
    a. molecular formula
    b. type formula
    c. structural formula

13. The reaction $Cl_2 + 2\,NaI \rightarrow 2NaCl + I_2 \uparrow$ is an example of chemical
    a. combination
    b. decomposition
    c. replacement

14. The reaction $P + O_2 \rightarrow P_2O_5$ is correctly balanced as
    a. $2P + O_2 \rightarrow P_2O_2$
    b. $4P + 5O_2 \rightarrow 2P_2O_5$
    c. $P_4 + 5O_2 \rightarrow 2P_2O_5$

15. The reaction $CaCO_3 \xrightarrow{\Delta} CaO + CO_2 \uparrow$ is an example of chemical
    a. combination
    b. decomposition
    c. replacement

1	
2	
3	
4	
5	
6	
7	
8	
9	
10	
11	
12	
13	
14	
15	

BASIC FACTS

Let us consider three principal classes of compounds: acids, bases, and salts, as well as water. In a *neutralization reaction*, an acid and a base react to form a salt and water.

Safety rule: always add acid to water, never the reverse.

PROPERTIES

Acids taste sour and turn blue litmus red.

Bases feel slimey and turn red litmus blue.

Salts are good electrolytes.

Water has a neutral taste. More substances dissolve in water than in any other known liquid.

Safety rule: Never taste a compound to determine its properties.

MEASUREMENT OF ACIDS AND BASES. The *pH (powers of hydrogen) scale*, showing the amount of available hydrogen (H^+) present expresses the acidity or alkalinity of a solution.

Order of Placement of Elements in Formulas for Compounds.
(1) If a compound has both metals and nonmetals, the metal stands first.
(2) If a compound has two nonmetals, carbon or hydrogen stands first and oxygen or chlorine come last.

NAMING COMPOUNDS. The following prefixes and suffixes are used in naming acids and salts:

Acid	Salt
hydro...ic	...ide
hypo...ous	hypo...ite
...ous	...ite
...ic	...ate
per......ic	per......ate

Hydro-ic acids and *-ide* salts are binary (two-element) compounds.

In the case of oxygen-containing acids and salts, the suffixes *-ous* and *-ite* are the lower oxidation state endings (less oxygen is therefore present); *-ic* and *-ate* are higher oxidation state (more oxygen) endings.

(Continued on page 48)

ADDITIONAL INFORMATION

An *acid* may be considered to be any compound that will give a hydrogen ion (H^+) in a solution. The positive ion part of this compound usually includes hydrogen. A *base* may be considered to be any compound that will give a hydroxyl ion (OH^-) in a solution. A *salt* is the product of the neutralization of an acid and a base. In addition to a salt, *water* (H_2O) is produced during the neutralization of an acid and a base.

Some Examples of Compounds:
Acids—HCl, HNO_3, H_2SO_4
Bases—$NaOH$, KOH, NH_4OH
Salts—$NaCl$, KNO_3, $(NH_4)_2SO_4$

Some Examples of Neutralization Reactions:
$HCl + NaOH \rightarrow NaCl + H_2O$
$HCl + KOH \rightarrow KCl + H_2O$
$HNO_3 + NaOH \rightarrow NaNO_3 + H_2O$

In different reactions, different salts are produced depending on what acids or bases are used. Water, however, is always formed in a neutralization reaction. Note that "salt" is a term for a chemical class. Table salt ($NaCl$), with which we are most familiar, is only one of many chemical salts.

The hydronium ion (or hydrated hydrogen ion) H_3O, rather than the H^+ ion, is actually formed when an acid dissociates in solution. The hydronium ion gives the sour taste to acids.

The *p*H number is the exponent of the hydrogen ion (H^+) concentration expressed in moles per liter. A *p*H of 7 is considered neutral, below 7 is acidic, above 7 is basic (alkaline).

$[H^+] = 0.1$ moles/l (10^{-1}),
$$pH = 1 \text{ (strong acid)}$$
$[H^+] = 0.0000001$ moles/l (10^{-7}),
$$pH = 7 \text{ (water)-Neutral}$$
$[H^+] = 0.00000000000001$ moles/l (10^{-14}),
$$pH = 14 \text{ (strong base)}$$

Naming Acids and Salts. As the number of oxygen atoms in a molecule increases, or as the positive oxidation state of the middle element in its formula increases, the prefixes and/or suffixes in the compound's name change.

The first term of the *salt* compound receives the name of the metal. The prefixes and suffixes are added to the second (nonmetal) term.

The oxidation state of the first term of the radical (the middle term of the compound) is more positive in acids where a greater amount of oxygen is present. The valence of the radical usually does not change.

(Continued on page 48)

EXPLANATIONS

1. Always add acid to water, never the reverse. The reverse produces more heat and might boil up in your face.

2. Red litmus turns blue at a pH above 7. Remember: *Base* → *Blue*

3. Blue litmus turns red at a pH below 7.

4. A pH of less than 7 is considered acidic.

5. A neutralization reaction is the reaction between an acid and a base.

6. A pH greater than 7 is considered basic.

7. The compound C_2H_6 is called ethane.

8. For example, Cu_2O is cuprous oxide and CuO is cupric oxide.

9. For example, HCl is hydrochloric acid.

10. Although both *-ite* and *-ate* are ternary endings, *-ate* is associated with a higher valence.

11. The compound may also be called iron III oxide. In this case iron is in its higher (trivalent) oxidation state.

12. This is a type formula for a whole series. If n (the number of carbon atoms in the molecule) is known, the molecular formula is readily calculated.

13. The addition of chlorine gas to the solution has displaced iodine from the compound and substituted chlorine. Free iodine is given off as a gas.

14. Phosphorus is balanced first ($2P + O_2 \rightarrow P_2O_5$). The least common multiple of 2 and 5 (the subscripts of oxygen) is 10($2P + 5O_2 \rightarrow 2P_2O_5$), but this leaves 4 phosphorus on the right side while only 2 on the left. Going back and changing the trial number of phosphorus on the left to 4 balances the equation.

15. This reaction is a chemical decomposition, or a breaking down of complex substances into two or more simpler ones. The triangle above the arrow ($\xrightarrow{\triangle}$) indicates that this reaction is heated. For every rise in temperature of 10° C, the rate of reaction will approximately double.

Answers

a	1
b	2
a	3
a	4
d	5
b	6
a	7
b	8
a	9
b	10
b	11
b	12
c	13
b	14
b	15

The *hypo-ous* acids have the element for which they are named present in a lower oxidation state than the *-ous* acids. The same is true for *hypo-ite* and *-ite* salts.

Per-ic marks the higher oxidation state use of *-ic* in naming acids. The same is true for *per-ate* and *-ate* salts.

A *molecular* formula (C_2H_6) shows the number of atoms of each element in the molecule.

An *empirical* formula (CH_3) shows the ratio of the number of atoms of one element in a molecule to the number of atoms of other elements.

A *type* formula (C_nH_{2n+2}) shows the subscript of one (or more) element-symbol(s) in terms of the subscript of the other element-symbol.

A *structural* formula (CH_3CH_3) shows how the molecules are linked together.

A *graphic* formula:

$$\begin{array}{c} \quad H \quad H \\ \quad | \quad \; | \\ H-C-C-H \\ \quad | \quad \; | \\ \quad H \quad H \end{array}$$

shows how the atoms are linked within the molecule.

An *electron dot* formula:

$$H:\ddot{C}\ddot{l}: \quad \text{(hydrochloric acid)}$$

KINDS OF CHEMICAL CHANGES. Four kinds take place: combination, decomposition, replacement, and double decomposition.

BALANCING EQUATIONS

For each element in an equation, there must be the same number of atoms on one side as on the other.

Complex equations are balanced by the *ion electron method*: balance the ions first, and then go back and balance the electrons.

Naming Oxides. When a *metal* forms two oxides, the endings, *-ous* and *-ic* are used.

When a *nonmetal* combines with oxygen alone to form two or more oxides, the *first term* of the name is the name of the element that is combining with oxygen, the prefix of the *second term* indicates the number of oxygen atoms in the molecule, and the suffix of the *second term* is the binary ending, *-oxide*. If the number of oxygen atoms is the same in two compounds (as in N_2O and NO), the valence endings *-ous* and *-ic* are used, and the second term has no prefix.

A common alternative method of naming chemical compounds such as oxides is the use of roman numerals to indicate the oxidation states of metals. For example, Cu_2O is copper I oxide, and CuO is copper II oxide.

Naming Bases. The hydroxyl radical (OH^-) is indicated by prefixing *hydro-* to the last term. Thus, for example, $Fe(OH)_2$ is named *ferrous hydroxide* (or iron II hydroxide), and $Fe(OH)_3$ is named *ferric hydroxide* (or iron III hydroxide).

KINDS OF CHEMICAL CHANGES. In *combination*, two or more substances form one complex substance: $2H_2 + O_2 \rightarrow 2H_2O$. In *decomposition*, a complex substance breaks down into two or more simpler ones: $CaCO_3 \xrightarrow{\Delta} CaO + CO_2$. In *replacement*, a more active element displaces a less active element from a compound: $Cl_2 + 2NaI \rightarrow 2NaCl + I_2$. In *double replacement* (double decomposition), an exchange of atoms or radicals takes place between compounds: $NaOH + HCl \rightarrow NaCl + H_2O$.

Balancing Simple Equations (Trial and Error Method). Start with a metal. Leave hydrogen to the next-to-last step, and leave oxygen to the last step. Remember that there is only *one* place in which you can put a number—in front of an element or a compound; never change or add subscripts. Balance the equation by using least common multiples. You may have to go back and change a trial number until you find one that works.

$$P + O_2 \rightarrow P_2O_5$$
$$2P + O_2 \rightarrow P_2O_5$$
$$2P + 5O_2 \rightarrow P_2O_5$$
$$4P + 5O_2 \rightarrow 2P_2O_5$$

(Also see Appendix III.)

13

TEMPERATURE, HEAT, GAS LAWS, AND THE MOLE CONCEPT

SELF-TEST

1. The heat needed to raise 1 g of water 1° C is 1
   a. calorie
   b. BTU
   c. specific heat

2. The heat needed to raise 1 lb of water 1° F is 1
   a. calorie
   b. BTU
   c. specific heat

3. The amount of heat needed to raise 1 g of oatmeal 1° C is its
   a. calorie
   b. BTU
   c. specific heat

4. If a substance goes from the solid state to the gaseous state at the same temperature, its heat of fusion + its heat of vaporization = its
   a. British thermal unit
   b. heat of sublimation
   c. specific heat

5. The weight of a substance times its change in temperature times its specific heat is expressed in units of
   a. heat
   b. state
   c. direction of heat flow

6. The product of the weight of a substance times its heat of fusion is used to calculate the heat necessary to change its
   a. temperature
   b. state
   c. direction of heat flow

7. 0° centigrade is equal to
   a. 0°F
   b. 460°R
   c. 273°K

8. The equation $A + B + C = ABC$ would symbolize
   a. Henry's law of solubility
   b. Dalton's law of partial pressures
   c. Graham's law of diffusion

9. The equation $S \propto P$ ($\propto$ means is proportional to) would symbolize
   a. Henry's law of solubility
   b. Dalton's law of partial pressures
   c. Graham's law of diffusion

10. The equation $\dfrac{r_1}{r_2} = \dfrac{\sqrt{d_2}}{\sqrt{d_1}}$ would symbolize
    a. Henry's law of solubility
    b. Dalton's law of partial pressures
    c. Graham's law of diffusion

11. The equation $V \propto T$ (P is constant) is an expression of
    a. Gay Lussac's law
    b. Charles' law
    c. Boyle's law

12. The equation $P \propto T$ (V is constant) is an expression of
    a. Gay-Lussac's law
    b. Charles' law
    c. Boyle's law

13. The equation $V \propto \dfrac{1}{P}$ (T is constant) is an expression of
    a. Gay-Lussac's law
    b. Charles' law
    c. Boyle's law

14. 7 l of gas at 0° C is chilled to −20° C. Its new volume is
    a. 7 l × 20
    b. 7 l × 250/273
    c. 7 l × 273/250

1 ___
2 ___
3 ___
4 ___
5 ___
6 ___
7 ___
8 ___
9 ___
10 ___
11 ___
12 ___
13 ___
14 ___

Temperature, Heat, Gas Laws, and the Mole Concept

BASIC FACTS

The two *critical points* at which a substance may change from one state of matter to another are the *fusion point* and the *vaporization point*.

Sublimation is a change between the solid and gaseous states, bypassing the liquid state.

One *calorie* is the amount of heat needed to raise the temperature of 1 gram of water 1°C.

One *BTU* is the amount of heat needed to raise 1 lb of water 1°F.

Specific heat is the number of calories needed to raise 1 gram of *any* substance 1°C, or the number of BTU's needed to raise 1 lb of any substance 1°F. Each substance has its own specific heat; and, for most substances, it is less than 1: ice = 0.51, water vapor = 0.482, Al = 0.217, Fe = 0.117, Cu = 0.092, glass (soda) = 0.16, and alcohol = 0.60.

Fahrenheit-Centigrade Conversion Factors:

$$°F = 9/5 °C + 32°$$
$$°C = 5/9 (°F - 32°)$$

In addition to the Fahrenheit and centigrade temperature scales in which both positive and negative figures appear, there are *absolute temperature scales* in which only positive figures appear.

The absolute temperature scales have *absolute zero* as their lowest measure. Absolute zero is that point at which so much heat is removed from a substance that the motion of molecules stops altogether.

Two absolute temperature scales are the *Rankine* (°R) and the *Kelvin*, or *absolute*, (°K, or °A) scales.

$$°R = 460° + °F \qquad °F = °R - 460°$$
$$°K = 273° + °C \qquad °C = °K - 277°$$

THE GAS LAWS. *Dalton's law* states that the total pressure $ABC =$

(Continued on page 52)

ADDITIONAL INFORMATION

The *fusion point* is that point at which a substance changes from a liquid to a solid (freezing point), or from a solid to a liquid (melting point). The *vaporization point* is that point at which a substance changes from a liquid to a gas (boiling point), or from a gas to a liquid (condensation point).

The change from dry ice (a solid) to CO_2 gas, or from CO_2 gas to dry ice, is an example of *sublimation*.

To change the state of a substance, *heat* (the energy of the motion of molecules) must be added or subtracted. The scientifically used *units of heat* are the calorie (4.18 joules) and the BTU (British thermal unit). Specific heat is a measure of the amount of heat necessary to raise the temperature of one gram of a substance 1°C. The *calorie* is a metric unit; the *BTU* is an English unit. Each substance has its own specific heat. The calorie and the BTU have water as the standard of measure.

WATER			
Heat of Fusion	Heat of Vaporization	Specific Heat	Heat of Sublimation
$80 \frac{cal}{g}$	$540 \frac{cal}{g}$	$1 \frac{cal}{(g)(°C)}$	$620 \frac{cal}{g}$

Heat of sublimation is heat of fusion plus heat of vaporization.

Heat units are used in calculating the amount of heat that must be added or subtracted to change a substance's state, temperature, or both. The amount of heat required to freeze 10 g of H_2O, for example, is

$$10 \, g \times 80 \frac{cal}{g} = 800 \, cal$$

A minus sign may be affixed to 800 calories to indicate that this amount of heat must be subtracted from the water in order to freeze it.

As heat is added to a substance, the molecular motion of the substance increases; as heat is subtracted, molecular motion decreases. *Absolute Zero* is that temperature at which all molecular motion ceases. Both *absolute temperature scales* developed from this concept. The *Kelvin scale* uses the units of the Centigrade scale: 100° between the freezing and boiling points of water. The *Rankine scale* uses the units of the Fahrenheit scale: 180° between the freezing and boiling points of water. The Centigrade and Kelvin scales are widely used in science today. The Rankine scale is not often used.

(Continued on page 52)

Temperature, Heat, Gas Laws, and the Mole Concept

EXPLANATIONS

1. By definition, a calorie is a unit in the metric system; 1000 calories equals one kilocalorie.

2. By definition, a BTU is a unit in the English system. One BTU equals 252 calories.

3. The heat needed to raise 1 g of any substance 1°C is defined as its specific heat.

4. Because the substance passes through both critical change-of-state points at the same temperature, there is no change in temperature.

5. Heat is usually measured in calories or in BTU's.

6. The heat of fusion is the heat required for one unit of the substance to change from a solid to a liquid.

7. The size of the units is the same in both systems. 0° K is the temperature at which all molecular motion is considered to stop.

8. Dalton's law of partial pressures states that the total pressure of a mixture of gases is the sum of the individual pressures of each gas.

9. Henry's law of solubility states that solubility is proportional to pressure.

10. Graham's law of diffusion states that rate of diffusion is inversely proportional to the square root of the mass density.

11. Charles' law states that volume is proportional to absolute temperature at constant pressure.

12. Gay-Lussac's law states that pressure is proportional to absolute temperature (at constant volume).

13. Boyle's law states that volume is inversely proportional to pressure (at constant temperature).

14. The temperature is falling, so a decreasing fraction is required. New volume = 71 × 250/273. (See below.)

Answers

a	1
b	2
c	3
b	4
a	5
b	6
c	7
b	8
a	9
c	10
b	11
a	12
c	13
b	14

THE TEMPERATURE-RELATED GAS LAW: *If temperature is changing, volume and pressure are directly proportional to it; if temperature is constant, volume is inversely proportional to pressure.* This law should be used with the following chart.

Property Fraction Chart	Changing Property	
	Rising	Falling
Temperature changing (proportion direct).	× increasing fraction	× decreasing fraction
Temperature constant, volume or pressure changing (proportion inverse).	× decreasing fraction	× increasing fraction

New Given Property = Old Given Property × Changing Property Fraction.

For Example: 5.0 l of gas @ 273°K is heated to 373°K. What is its new volume?

As the temperature is rising, an increasing fraction is needed (larger number on top). New volume = 5.0 l × 373/273 = 6.8 l.

$A + B + C$, where A, B, and C are partial pressures.

Henry's law states that $S \propto P$, where S is solubility and P is pressure.

Graham's law states that $\dfrac{r_1}{r_2} = \dfrac{\sqrt{MW_2}}{\sqrt{MW_1}} = \dfrac{\sqrt{d_2}}{\sqrt{d_1}}$, where r is a rate of diffusion.

Gay-Lussac's law states that when volume is constant, $P \propto T$.

$$\dfrac{P_1}{P_2} = \dfrac{T_1}{T_2} \text{ or } P_2 = P_1 \dfrac{T_2}{T_1}$$

Charles' law states that, when pressure is contant, $V \propto T$.

$$\dfrac{V_1}{V_2} = \dfrac{T_1}{T_2} \text{ or } V_2 = V_1 \dfrac{T_2}{T_1}$$

Boyle's law states that, constant temperature $V \propto \dfrac{1}{P}$

$$\dfrac{P_1}{P_2} = \dfrac{V_2}{V_1} \text{ or } P_2 = P_1 \dfrac{V_1}{V_2} \text{ or } V_2 = V_1 \dfrac{P_1}{P_2}$$

STP (standard temperature and pressure) is defined as 0°C and 760 mm of mercury, or 273°K and 760 mm Hg.

The temperature-related gas law links the pressure, volume, and temperature. It may be expressed as:

$$PV = nRT$$

where n is the number of moles of gas involved and R is a constant. Its effects may be most simply grasped by the procedure explained on the Explanations page.

The concentration of solutions is expressed as percent composition, as molarity (number of moles per liter of solution), as molality (number of moles per 1000 grams of solution), or as normality (number of equivalents per liter of solution).

The *mole* is defined as one gram molecular weight (GMW) of any element. The mole concept stems from *Avagadro's hypothesis* that equal volumes of gas, at the same temperature and pressure, have the same number of molecules.

	°F	°R	°C	°K
Boiling Point of Water	212	672	100	373
Room Temperature (approximately)	62.6	522	17	290
Freezing Point of Water	32	492	0	273
	0	460		
Oxygen Liquifies	−298	162	−183	90
Hydrogen Liquifies	−424	36	−253	20
He$_4$ Liquifies	−452.4	7.6	−268.8	4.2
He$_3$ Liquifies	−453.3	6.7	−269.3	3.7
Absolute Zero	−460	0	−273	0

Henry's law of solubility states that the solubility, by weight, of a gas in a liquid is proportional to the pressure applied.

Graham's law of diffusion states that the rates of diffusion of different gases are inversely proportional to the square roots of their molecular weights, or to the square roots of their densities.

Gay-Lussac's law states that when the volume remains constant, the pressure of a gas is directly proportional to its absolute temperature.

Charles' law states that when the pressure remains constant, the volume of a gas is directly proportional to its absolute temperature.

Boyle's law states that when the temperature remains constant, the volume of a gas is inversely proportional to its pressure.

Avagadro's hypothesis has been confirmed experimentally, and it has been shown that one GMW of any gas at STP will occupy 22.4 liters (l) and contain 6.02×10^{23} molecules.

Percent Composition (parts per 100)

$$= \text{the part} \div \text{the whole} \times 100.$$

The *mole* is a chemical device for readily converting between weight, volume, and number of molecules per sample.

$$\text{Molarity (M)} = \dfrac{\text{Moles}}{\text{liter}} \qquad \text{Normality} = \dfrac{\text{equivalents}}{\text{liter}}$$

14 ORGANIC CHEMISTRY
SELF-TEST

1. A compound of only hydrogen and carbon is called a(an) _____ .

2. The group of hydrocarbons characterized by straight (or branched) chains of carbon atoms is the _____ hydrocarbons.

3. If any other element replaces a hydrogen in a hydrocarbon, the new compound is called a(an) _____ .

4. The name of the compound with a molecular formula of C_8H_{18} is _____ .

5. The alkane with ten carbon atoms would have _____ hydrogen atoms.

6. An aliphatic compound on which a hydroxyl radical has replaced a hydrogen is called a(an) _____ .

7. The compound C_2H_5OH is called _____ .

8. The alcohols are the organic equivalents of _____ .

9. Compounds in which two alkyl groups are joined together by an oxygen are called _____ .

10. The esters are the organic equivalents of _____ .

11. Organic compounds in which hydrogen and oxygen are present in a 2:1 ratio are called _____ .

12. Complex organic molecules that build body tissue are called _____ .

13. Organic catalyst compounds are called _____ .

14. Chemical messenger compounds are called _____ .

15. Organic compounds that control reproduction are called _____ .

16. Unsaturated compounds are characterized by the presence of _____ bonds between carbon atoms.

17. Compounds with the same molecular formula but different graphic formulas are called _____ .

18. Aromatic series compounds are _____ compounds, rather than straight, or branched, chain compounds.

19. The "starting point" of the aromatic series is the compound _____ , whose molecular formula is C_6H_6.

20. Simple sugars are termed _____ .

1 _____
2 _____
3 _____
4 _____
5 _____
6 _____
7 _____
8 _____
9 _____
10 _____
11 _____
12 _____
13 _____
14 _____
15 _____
16 _____
17 _____
18 _____
19 _____
20 _____

BASIC FACTS

Organic chemistry is essentially the chemistry of the compounds of carbon.

Two basic kinds of carbon compounds exist: hydrocarbons and derivatives. A *hydrocarbon* is a compound containing only the elements hydrogen and carbon (although the term is sometimes used rather loosely). A *derivative* is formed when any other element or elements are added to a hydrocarbon.

Organic compounds occur in long series of related compounds. The principal group of hydrocarbons is the *aliphatic hydrocarbons* (or *paraffins*), made up of various chainlike series of hydrocarbons.

The members of the simplest aliphatic series, the alkanes, are found in various petroleum fractions.

Double bonds and *triple bonds* are formed between atoms (usually carbon) where adjacent hydrogen atoms are removed from a molecule. Compounds with double or triple bonds between carbon atoms are termed unsaturated. Members of the *alkane* series (type formula: C_nH_{2n+2}) are saturated. Members of the alkenes (C_nH_{2n}) and the alkynes (C_nH_{2n-2}) are unsaturated. The names of alkanes end in *-ane*, of alkenes in *-ene* and of alkynes, in *-yne*. The reactions of the alkanes are used here to typify the reactions of all the aliphatic hydrocarbons. Note that the *valences* of the atoms (C = 4, O = 2, H = 1) *must be satisfied* in all cases by the appropriate number of (covalent) bonds.

Isomers are compounds with the same molecular formula (same number of atoms of each element) but different graphic formulas.

(Continued on page 56)

ADDITIONAL INFORMATION

Organic chemistry is defined as the chemistry of carbon and its compounds. The term organic chemistry stems from the association of carbon compounds with living matter. At one time such compounds were thought to be found only in nature, but most are now made in the laboratory.

The term *aliphatic* means fat-like or oily. *Paraffin* means "little affinity" and refers to its chemical inactivity. Though a paraffin will burn, it does not readily unite with other atoms or molecules until a hydrogen atom or two is removed, creating a *radical* with a free valence bond. It is this radical that is chemically active and forms derivatives.

Hydrocarbons of the Alkane Series

Name	Molecular Formula	Name	Molecular Formula
Methane	CH_4	Hexane	C_6H_{14}
Ethane	C_2H_6	Heptane	C_7H_{16}
Propane	C_3H_8	Octane	C_8H_{18}
Butane	C_4H_{10}	Nonane	C_9H_{20}
Pentane	C_5H_{12}	Decane	$C_{10}H_{22}$

Note that from 5 carbon atoms on (in the molecular formula) the names are derived from Latin numerals.

Methane is the "marsh gas" given off by stagnant decaying vegetation. It is a prime fuel in natural gas from wells.

Liquified Petroleum Gases (LPG) are fuel gases (ethane, propane, and butane) cooled to a liquid for safer transport and storage. "Gasoline" is a mixture, or blend, of several of the C_5 to C_{10} aliphatic hydrocarbons and their derivatives.

The name of a compound in which there are two double bonds present will end in *-diene*. The position of the double bonds (and of substituent groups in derivatives) is indicated by numbers before the name, when necessary. For example, $CH_2=CH-CH=CH_2$ is 1,3-butadiene.

Gentle oxidation of a hydrocarbon may involve the addition of oxygen or the removal of hydrogen. Removal of hydrogen from a compound results from a hydrogen combining with free excess oxygen to form water as a by product. The hydrogen-oxygen bond is stronger than the hydrogen-carbon bond. An *alkyl radical* which, like other radicals has a free bond for attachment, is formed by removing a hydrogen from a hydrocarbon.

Alcohols are formed by the gentle oxidation of an alkane or the fermentation of sugars. Methyl (industrial, or wood) alcohol is used as antifreeze in car radiators. Ethyl (grain) alcohol is used medicinally and is the alcohol present in "alcoholic beverages." A secondary alcohol is an alcohol whose hydroxyl group is attached to a carbon other than an end carbon in the chain.

(Continued on page 56)

Organic Chemistry

EXPLANATIONS

1. Hydrocarbons are either aliphatic (chain structure) or aromatic (ring structure).

2. Most of our petroleum and plastic products start with the aliphatic hydrocarbons.

3. A hydrocarbon to which any other element has been added is a derivative.

4. Octane is an alkane.

5. Use the type molecular formula C_nH_{2n+2}. If $C = 10$, $H = 2(10) + 2 = 22$.

6. Alcohols are formed by the gentle oxidation of an alkane or the fermentation of sugar.

7. Ethyl alcohol (ethanol) is also called grain alcohol.

8. Both have an —OH functional group.

9. Ethers are a type of organic oxide.

10. Organic bases (alcohols) + carboxylic acids → organic salts (esters) + water.

11. Carbohydrates include sugars, starches, and celluloses.

12. Proteins are complex amino acid chains.

13. Without enzymes, life would be impossible.

14. Hormones regulate a number of body processes.

15. The principal nucleic acids are deoxyribonucleic acid (DNA) and ribonucleic acid (RNA).

16. The alkenes are chain compounds with unsaturated double bonds. The alkynes are chain compounds with unsaturated triple bonds.

17. Isomers differ in the spatial arrangement of their atoms.

18. There is a double bond between every other carbon in aromatic rings.

19. The graphic formula for benzene is:

20. Monosaccharides have the molecular formula $C_6H_{12}O_6$. There are also disaccharides ($C_{12}H_{22}O_{11}$) and polysaccharides (($C_6H_{10}O_5)_n$).

Answers

hydrocarbon	1
aliphatic	2
derivative	3
octane	4
22	5
alcohol	6
ethyl alcohol	7
bases	8
ethers	9
salts	10
carbohydrates	11
proteins	12
enzymes	13
hormones	14
nucleic acids	15
double or triple	16
isomers	17
ring	18
benzene	19
monosaccharides	20

	SOME CHAIN DERIVATIVES OF THE ALKANE SERIES			
	Hydrocarbons		Derivatives	
	Alkane	Alcohol	Aldehyde	Carboxylic Acid
Structural Formula	C_2H_6	C_2H_5OH	CH_3CHO	CH_3COOH
Graphic Formula	H—C—C—H (with H's)	H—C—C—O—H (with H's)	H—C—C=O (with H's)	H—C—C=O / O—H
Classic Name	ethane	ethyl alcohol	acetaldehyde	acetic acid
New Name		ethanol	ethanal	ethanoic acid

Derivatives are the organic equivalents of inorganic compounds. Alcohols, aldehydes, ethers, ketones, carboxylic acids, and esters are six of the families of hydrocarbon derivatives. Alcohols are organic bases (—OH group present). Ethers consist of two organic groups joined together by an oxygen. Ketones are organic carbonates ($-\overset{\overset{\text{O}}{\|}}{\text{C}}-$ group present). Carboxylic acids have the $-\text{C}{\overset{\diagup\text{O}}{\diagdown\text{OH}}}$ group present. Esters are produced by the reaction of organic acids with alcohols. They consist of two organic groups joined by a $-\text{C}{\overset{\diagup\text{O}}{\diagdown\text{O}-}}$ group.

The *aromatic* hydrocarbons are the other major group of organic compounds. They are named for their associated odors. These are *ring compounds* rather than straight or branched chains (as the aliphatics are). There are double bonds between alternate carbons. Each carbon is attached to other carbons in a ring with hydrogen and other substituents attached to the outside of the ring.

Most of the materials of LIVING ORGANISMS are organic chemicals such as *carbohydrates, fats, proteins, enzymes, hormones, vitamins,* and *nucleic acids.*

The *sugars* are the most readily used form of stored chemical energy; the *fats* are the most concentrated forms. Fats are the esters made by glycerine reacting with fatty acids. *Proteins* are complex amino acid chains. They serve to build body tissue. *Enzymes* are organic catalysts. *Hormones* are chemical messengers. *Vitamins* are disease restricting co-enzymes. *Nucleic acids* control reproduction.

The *aldehydes* are formed by a mild oxidation of alcohols. Addition of free excess oxygen pulls two hydrogen atoms out of the alcohol to form water, producing the aldehyde. Formaldehyde is used as a disinfectant and as embalming fluid.

Mild oxidation of aldehydes, or more vigorous oxidation of alcohols, produces the *carboxylic acids*. Formic acid is injected by the bite of the fire ant, and acetic acid is the mild acid in vinegar.

The *ethers* are formed by the dehydration of (removal of water from) alcohols. The ether made from ethane is the familiar hospital anesthetic. Ketones are the product of the gentle oxidation of a secondary alcohol. Acetone (dimethyl ketone) is an industrial solvent for plastic resins, paint, etc. The *esters* are formed by combining a carboxylic acid with an alcohol (which is the organic equivalent of a base since it has a hydroxyl radical) in a reaction analogous to a neutralization. Ethyl acetate is an ester familiar as nail polish remover.

Amines are derivatives in which hydrogen is replaced by an NH_2 group. *Amino acids* are compounds in which both carboxyl groups and amines are present. They are important in the chemistry of living organisms.

Benzene (C_6H_6) is the "starting point" of the aromatic series. Every other carbon-to-carbon bond in the benzene ring may be considered a double bond, thus satisfying the valence requirements for carbon and hydrogen. (Actually, the three double bonds are spread over the entire molecule, thus giving an effective one-and-one-half bonds between carbons.)

Carbohydrates are compounds composed of carbon, hydrogen, and oxygen; the hydrogen and oxygen being present in a ratio of 2 to 1. The sugars are the most typical of the carbohydrates.

The principal nucleic acids are DNA and RNA. *DNA* (deoxyribonucleic acid) exists in the nucleus of a living cell and controls the cell's heredity. It is the sole molecule capable of reproducing itself. *RNA* (ribonucleic acid) molecules move as "messengers" from the nucleus into the cell, and build protein molecules from amino acids.

15 RELATIVITY AND QUANTUM MECHANICS

SELF-TEST

1. The Michelson-Morley experiments
   a. disproved "ether"
   b. discovered X-rays
   c. showed light to be both a particle and a wave

2. Planck's Hypothesis
   a. disproved "ether"
   b. discovered X-rays
   c. showed light to be both a particle and and a wave

3. Rutherford
   a. anticipated relativity
   b. developed the Theory of Relativity.
   c. established the modern concept of the atom

4. Einstein
   a. anticipated relativity
   b. developed the Theory of Relativity
   c. established the modern concept of the atom

5. Fitzgerald and Lorentz
   a. anticipated relativity
   b. wrote the Theory of Relativity
   c. explained radioactivity

6. The Special Theory of Relativity deals with
   a. uniform motion
   b. accelerated motion
   c. circular motion

7. The General Theory of Relativity deals with
   a. accelerated motion
   b. circular motion
   c. both a and b

8. Bohr
   a. explained X-ray frequency loss
   b. described a hydrogen atom
   c. showed the wave nature of all small particles

9. De Broglie
   a. explained X-ray frequency loss
   b. described a hydrogen atom
   c. showed the wave nature of all small particles

10. Position and velocity cannot be determined in the same experiment
    a. states the Uncertainity Principle
    b. states the Exclusion Principle
    c. explains Quantum Mechanics

11. Schrödinger and Heisenberg
    a. formed the Uncertainity Principle
    b. formed the Exclusion Principle
    c. developed Quantum Mechanics

12. "No two atomic particles can have the same set of properties"
    a. states the Uncertainity Principle
    b. states the Exclusion Principle
    c. explains Quantum Mechanics

13. The *m* quantum number tells
    a. which shell an electron is in
    b. direction of electron revolution
    c. direction of electron spin

1
2
3
4
5
6
7
8
9
10
11
12
13

BASIC FACTS

PRELUDES TO RELATIVITY.
Michelson-Morley Experiment (1887).
Fitzgerald-Lorentz Contraction (1893).
Planck's Quantum Hypothesis (1900).

SPECIAL THEORY OF RELATIVITY (1905). (1) The laws of nature appear the same to all observers in uniform motion relative to one another.

PRINCIPAL COROLLARY: The speed of light in a vacuum is invariable. (2) Time must be included in the frame of reference (dimensions) of each observer.

SPECIAL CONSEQUENCES: Length, mass, and time are proportional to speed of movement. As the speed of an object approaches the speed of light, its length decreases, mass increases, and time measurement shortens.

COROLLARY: Mass and kinetic energy are equivalent.

$$E = mc^2$$

GENERAL THEORY OF RELATIVITY. The general theory deals with accelerated and circular motion. Gravitation (as the principal factor in these motions) is shown to be equivalent to the force of inertia. Einstein proposed that a large mass creates a gravitational field around itself. This supplanted Newton's view of a gravitational force that acts over vast distances instantly.

Einstein mathematically predicted the following astronomical phenomena which were subsequently verified. (1) Advance of planetary perihelion (verified for Mercury, Venus, and Earth). (2) Red shift of spectra due to gravitation. (3) Deflection of starlight near a large mass.

(Continued on page 60)

ADDITIONAL INFORMATION

The *Michelson-Morley* experiment attempted to determine the orbital velocity of the earth by measuring the speed of light both in the direction of the earth's motion, and at right angles to it. It failed to find any differences, arriving at the same speed of light in all directions. This destroyed the early notions of "ether" and "absolute motion."

The *Fitzgerald-Lorentz Length Contraction Theory* assumed a decrease in length by a factor $\sqrt{1 - v^2/c^2}$, in the direction of motion, as the speed of an object increases. (Lorentz said mass increases with speed.) This was suggested to explain the failure of the Michelson-Morley experiment in terms of shortened arm length on the part of the measuring instruments. However, using arms of unequal length (which should therefore be shortened by different amounts), produced the same results.

Planck's Quantum Hypothesis proposed that light is emitted only in whole-number packets or quanta of photons. The energy of the quanta is proportional to their frequency, $E = hf$, where h is Planck's constant (5.62×10^{-27} erg-sec).

SPECIAL THEORY OF RELATIVITY. The *speed of light* in a vacuum is the same, regardless of the motion of the observer or the source, for all observers moving uniformly relative to one another. This followed from the results of the Michelson-Morley experiment.

An object is only described adequately by using four dimensions of length, width, height, and *time*. Any description (or mathematical treatment) must deal with a four-dimensional space-time continuum.

The *length decrease* and *mass increase* equations of the essentially mathematical Special Theory incorporated the work of Fitzgerald and Lorenz into a larger framework. The corollary of the *equivalence of mass and energy* was the theoretical basis for the development of the atomic bomb.

That mass increase does occur has been verified by the resultant slowing of nuclear particles at high speeds in particle accelerators.

Time dilation is the slowing down of time for an observer in motion relative to the time noted by a stationary observer. Thus, a man traveling at a speed near the speed of light might age only one year, but return to the earth to find his family 20 years older.

A heated metal filament in a vacuum tube gives off electrons. This is known as the *Edison effect*. A positive metal plate at the other end will collect these, providing an electric current in its simplest form—a current without a wire. This device is called a *cathode ray tube*. If an electron moving in this tube abruptly hits a heavy metal block, a very penetrating wave is deflected off. *Roentgen* called these *X-rays* because he didn't know what they were.

(Continued on page 60)

EXPLANATIONS

1. The Michelson-Morley experiment disproved "ether" by proving the speed of light in a vacuum was constant.

2. Planck's Hypothesis states light was both a particle and a wave, emitted in whole-number packets.

3. Rutherford established the modern concept of the atom as a dense positive nucleus, surrounded at great distances (relative to nuclear size) by negative orbiting electrons.

4. Einstein formulated the Theory of Relativity, based on the prior work of Plank, Fitzgerald, Lorentz and others.

5. Fitzgerald said length decreases with high speed. Lorentz said mass increases with high speed.

6. The Special Theory of Relativity deals with uniform motion, which is straight-line and non-accelerated.

7. The General Theory of Relativity deals with both circular and accelerated motion.

8. Bohr's model is still used for simple atoms.

9. This duality of matter is a basis of modern physics.

10. That position and velocity cannot be determined in the same experiment is a statement of Heisenberg's Uncertainty Principle.

11. Schrödinger and Heisenberg, with their "wave mechanics" and "matrix mechanics" respectively, developed quantum mechanics.

12. That two atomic particles can have the same set of properties is a statement of Pauli's Exclusion Principle.

13. The m quantum number tells direction of electron revolution.

Answers

a	1
c	2
c	3
b	4
a	5
a	6
c	7
b	8
c	9
a	10
c	11
b	12
b	13

60 Relativity and Quantum Mechanics

PRELUDES TO QUANTUM MECHANICS.

The Edison Effect	(1883)
Roentgen's Discovery of X-Rays	(1895) (1895)
Einstein's Photoelectric Effect	(1905)
Rutherford's Modern Atom Concept	(1911)
Bohr's Hydrogen Model Atom	(1913)
DeBroglie's Matter Waves	(1923)
The Compton Effect	(1923)
Pauli's Exclusion Principle	(1925)

QUANTUM MECHANICS (1925) is a strictly mathematical approach to a workable description of atomic particles. The principles were developed independently by Schrödinger and Heisenberg. Schrödinger's *"wave mechanics"* treats an electron as a wave and gives a model that can be visualized. Heisenberg's *"matrix mechanics"* gives an abstract mathematical interpretation.

Quantum numbers are used to describe the possible value of observable properties of atomic particles. One number (n) tells which shell on electron is in (how far out from the nucleus is the path in which it moves). Another number (l) tells which subshell it is in (the shape of its orbit). A third (m) tells the direction of electron revolution. A fourth (s) gives the direction of electron spin. These numbers adequately describe a particle.

The *Exclusion Principle* states that no two electrons in an atom can have the same set of quantum numbers.

Heisenberg's *Uncertainty Principle* states that motion and position cannot both be determined in the same experiment on a small particle.

Einstein explained the giving off of electrons from some light metals when light strikes the surface (the *photoelectric effect*) in terms of quanta. A curious effect was known. Increasing intensity (from dim to bright) caused more electrons to be emitted, but they all had the same speed. Reducing the frequency of the light reduced electron speed but below a specific frequency no electrons were emitted. Einstein proposed that a certain minimum amount of energy was needed to break the bond binding the electron to the nucleus. The energy of the low frequency light did not exceed this, so no electrons emerged. Higher energies (frequencies) served to increase electron speed, after the bond was broken. Photon energy − bond energy = energy of movement (or $hf - w = \frac{1}{2}mv^2$). X-ray production agreed with this, since faster electrons produced more penetrating X-rays.

Rutherford, after observing deflection of alpha particles from very thin foil, conceived and published the essentially modern concept of the atom as a dense, positive nucleus surrounded at (relatively) great distances by negative orbiting electrons.

Electromagnetic theory demands that a moving electron should radiate energy, lose velocity, and eventually be drawn into the nucleus. At certain energy levels, it doesn't do this, however. In his *atomic model*, Bohr applied quantum theory to this phenomenon. An electron can absorb only those quanta whose energy will send it from its normal energy level to a higher energy level corresponding to a whole number of wavelengths. When returning from this *excited state*, the electron releases energy as electromagnetic radiation. It may return by a single large jump or by a series of smaller ones, thus producing many *spectral lines* instead of just one.

In *fluorescence*, ultraviolet light sends the electron to the excited-state level. Coming back, it radiates in the lower-frequency visible light range. If visible emission persists after the ultraviolet light is turned off, it is called *phosphorescence*.

De Broglie showed that, just as light waves have a particle nature, all (extremely small) particles have a wave nature. The frequency of *matter waves* is proportional to mass times velocity divided by Planck's constant.

$$f = \frac{mv}{h}$$

Compton explained the X-ray frequency reduction effect after impact and deflection in terms of quanta. (Photon energy$_1$ − electron movement = photon energy$_2$.)

The Pauli Exclusion Principle states that no two like particles in an atom could have the same set of properties.

16 RADIOACTIVITY

SELF-TEST

1. The radioactivity of uranium was discovered by _____.

2. The radioactivity of uranium was shown to be of three types by _____.

3. Identify: the nucleus of the helium atom.

4. Identify: a negative electron, emitted from the nucleus.

5. The spontaneous emission of particles and rays by unstable nuclei is called _____.

6. The elements in the Periodic Table with an atomic number greater than _____ are radioactive.

7. The length of time required for one-half the weight of a given material in a radioactive series to change into the material of the next step is called _____.

8. A straight-line accelerator is called a(an) _____.

9. An accelerator that spirals a positive particle into an ever-expanding spiral before releasing it is called a(an) _____.

10. An accelerator of the type described in question 9 that accelerates negative beta particles is called a(an) _____.

11. The kinetic energy given an electron by a difference of potential of 1 volt is _____.

12. The product of combination of a positive and negative electron pair is a _____.

13. After ejection of an alpha particle, atomic number _____.

14. After ejection of a beta particle, atomic number _____.

1 _____
2 _____
3 _____
4 _____
5 _____
6 _____
7 _____
8 _____
9 _____
10 _____
11 _____
12 _____
13 _____
14 _____

BASIC FACTS

RADIOACTIVITY—DISCOVERY AND INVESTIGATION.

Roentgen Discovers X-Rays	(1895).
Becquerel Discovers Radioactivity	(1896).
Curies Discover Radium	(1898).
Rutherford Distinguishes α, β, and γ Rays	(1899).
Rutherford Explains Radioactivity	(1903).

Radioactivity is the spontaneous emission of alpha, beta, and gamma rays from the unstable nuclei of certain atoms. *Rutherford* showed the radioactivity of uranium to be of the three types: alpha particles (two protons and two neutrons—helium nucleus), beta particles (a negative electron that emerges from the nucleus), and gamma rays (an electromagnetic wave).

Natural Radioactive Elements. Any element above bismuth (atomic number 83) is radioactive. There are several series of radioactive elements centering around uranium that naturally decay to end points centering around lead. There may be as many as 16 steps in these series. Geologists use these series as atomic clocks to date the rocks of the earth. A few other single-step series are important—primarily the decay of carbon-14. Carbon-14, which is present in all living things, is useful to the archeologist as it may be used to date material back about 40,000 years.

ARTIFICIAL RADIOACTIVE ELEMENTS may be created by bombardment of the atom's nucleus with high-speed particles. Atomic particle accelerators are used to do this. Charged particles may be accelerated by high-voltage machines to energies of MeV's (million electron volts) or BeV's (billion electron volts).

The more common particles are used in the accelerators to produce

(Continued on page 64)

ADDITIONAL INFORMATION

Roentgen named his mysterious rays X-rays, after the symbol in algebra for an unknown quantity, because he didn't know what they were. Becquerel accidentally discovered that some sort of emission from uranium ore had exposed a photographic plate. The Curies were set to work on uranium by Becquerel to find out what was causing the emission. After two years of chemically sorting through several tons of material, they came up with a fraction of a gram of a substance which they named radium.

The discovery of radium prompted much experimentation. Rutherford put uranium ore between two magnets and saw three types of ray paths resulting. What he learned was approximately as follows:

	Charge	Speed	Penetration (stopped by)
alpha α	positive	slowest	a few sheets of paper
beta β	negative	faster	a few mm of lead
gamma γ	none	fastest	a few in. of lead

The Beta particle is not an ordinary orbiting electron. It is ejected from the nucleus at great velocity when a neutron decays into a proton and a beta particle. The proton remains in the nucleus. Also ejected is a *neutrino*, which has zero mass when at rest. In some instances, the beta particle emerges with most of the energy of the pair. In other cases, the neutrino carries most of the kinetic energy. Thus, successive beta particles from the same nucleus may have different velocities.

EFFECTS OF RADIOACTIVE PARTICLE EMISSION. Loss of an alpha particle decreases atomic number by 2, and atomic weight by 4. Loss of a beta particle increases atomic number by 1; atomic weight remains approximately the same. Loss of a gamma ray, which often occurs after alpha or beta particle emission, causes no change in atomic number or atomic weight. The gamma ray has the shortest wave length of the electromagnetic spectrum.

RADIOACTIVE SERIES. The average for the lengths of time between the decay product steps for a large number of atoms are definite and precise, but by no means equal. Emission is from the nucleus and entirely independent of any outside condition. The unit of measure is the half-life. The *half-life* is the length of time required for one-half of the weight of a given material in a radioactive series to change into the material of the next step.

KINDS OF ACCELERATORS. In a *linear accelerator*, a particle sent down the tube is "drawn" still faster by a mag-

(Continued on page 64)

EXPLANATIONS

1. The radioactivity of uranium was accidentally discovered by Becquerel.

2. Rutherford showed the radioactivity of uranium to be composed of alpha, beta, and gamma rays.

3. The nucleus of the helium atom is composed of two neutrons and two protons.

4. A negative electron from the nucleus is a beta particle.

5. Radioactive isotopes of elements that are not naturally radioactive can be made.

6. All elements beyond bismuth in the Periodic Table are naturally radioactive.

7. Four grams of an isotope with half-life of one year would decay to 2 g in one year, 1 g in two years, and so forth.

8. Linear accelerators may be very long.

9. Any accelerator that cycles or spirals a positive particle is a cyclotron.

10. Such an accelerator that accelerates negative beta particles is called a betatron.

11. The kinetic energy given 1 electron by a difference of potential of 1 volt is defined as 1 electron volt.

12. Matter and antimatter annihilate each other.

13. After ejection of an alpha particle, atomic number decreases by 2, and atomic weight decreases by 4.

14. After ejection of a beta particle, atomic number increases by 1; atomic weight is virtually unchanged.

Answers

Answer	#
Becquerel	1
Rutherford	2
alpha particle	3
beta particle	4
radioactivity	5
bismuth	6
half-life	7
linear accelerator	8
cyclotron	9
betatron	10
electron volt	11
gamma ray	12
decreases by 2	13
increases by 1	14

Element	U	Th	Pa	U	Th	Ra	Rn	Po	Pb	Bi	Po	Pb	Bi	Po	Pb
atomic number	92	90	91	92	90	88	86	84	82	83	84	82	83	84	82
atomic weight	238	234	234	234	230	226	222	218	214	214	214	210	210	210	206
emission	α	β	β	α	α	α	α	α	β	β	α	β	β	α	

Radioactive Decay Series of Uranium-238

bombardment. Other nuclear particles are produced by these bombardments. The neutron, having no charge, will easily penetrate to the nucleus. The proton, having a positive charge, must be greatly accelerated to penetrate. The electron, being much less massive, must move even faster to have any effect on the nucleus at all.

OTHER NUCLEAR PARTICLES. A great many atomic particles are known. All but one (the orbiting electron) come from the nucleus. Most are seen coming from the nucleus after bombardment by neutrons, protons, alpha particles, beta particles, or cosmic rays.

Almost every particle known seems to have an *antiparticle* of opposite charge. The electron is matched by the positive positron. The proton has an antiproton and even an antineutron exists. Matter may exist in two forms, then, one of them being antimatter.

Some Other Nuclear Particles

- μ-MESONS (MUONS)—positively or negatively charged particles having a mass (207/1837 amu) falling between that of an electron and that of a nucleon (proton or neutron).
- π-MESONS (PIONS)—various positively or negatively charged (or neutral) particles having masses somewhat greater than a muon's, but less than a nucleon's (264/1837-273/1837 amu).
- K-MESONS—uncharged particles with masses between those of muons and those of nucleons.
- HYPERONS (V PARTICLES)—particles slightly heavier than nucleons. Positive and negative hyperons have masses of about 1.21 amu. There are neutral hyperons with masses of 1.20 and 1.46 amu.
- LEPTONS—PHOTONS and NEUTRINOS (particles without charge or mass).

---

netic field. At the proper moment, the field is reversed. This gives the particle a "kick in the pants" that speeds it along with even more velocity towards the next field. However, since the particle is now moving even faster, this means that each successive field must be farther and farther "downstream." Linear accelerators therefore tend to grow lengthy. Stanford University is building one almost 2 miles long (20 BeV).

A *cyclotron* uses two huge D-shaped electromagnets (◖ ◗). As positive particles follow a circular path in a magnetic field and a straight line between the magnets, they follow a spiral path in a cyclotron. Thus, the path is kept long, but the length of the machine is greatly reduced. Brookhaven National Laboratory near New York City has a 33 BeV cyclotron. (A *betatron* accelerates beta particles, and it is thus necessary to reverse the polarity of the electromagnets.)

Neutrons are already going faster than is generally desired. To slow them down to thermal neutrons (neutrons traveling at room temperature speeds), heavy water, graphite, or cadmium is used.

Antiparticles are often produced by nuclear bombardment. In some experiments, a particle (the *positron*) having the mass of an electron but an opposite charge has been found as a product of artificial disintegration and with cosmic ray phenomena. When positive and negative electrons combine, they disappear and a gamma ray or photon is produced. Conversely, a gamma ray passing near a charged body (e.g. an atomic nucleus) may break down into a positive and negative electron pair (pair production).

We are familiar with an orderly world of atoms composed of positive protons and neutral neutrons in the nucleus, and negative electrons in orbit around it. What is to prevent an atom being formed of negative protons and neutral neutrons in the nucleus, and positive electrons in orbit around it? There seems to be no good reason why such *antimatter* could not possibly exist.

17 INTERCONVERSION OF MASS AND ENERGY

SELF-TEST

1. A large, heavy atom splitting into smaller atoms with release of energy is atomic _____.

2. The fissionable uranium ore used in the atomic bomb is uranium _____.

3. One neutron striking a fissionable uranium-235 atom *may* give as many as _____ neutrons.

4. If insufficient fissionable material is present to produce a chain reaction, the mass is _____.

5. If each fission produces an average of one new fission, the mass is _____.

6. If each fission produces two or three new fissions, the mass is _____.

7. The part of an atomic reactor that surrounds the fuel rods to slow down the rate of neutron travel is the _____.

8. Between the fuel rods are the _____, which absorb neutrons.

9. Two subcritical masses that are suddenly blown together to produce a supercritical mass form a(an) _____.

10. A process where smaller atoms combine to form a larger atom plus much energy is _____.

11. The fourth state of matter is _____.

12. One gram of matter converted entirely to energy equals the energy released by the explosion of _____ tons of TNT.

13. An atomic reaction in which the number of free bombarding neutrons increases extremely rapidly is termed a(an) _____ reaction.

1 _____
2 _____
3 _____
4 _____
5 _____
6 _____
7 _____
8 _____
9 _____
10 _____
11 _____
12 _____
13 _____

BASIC FACTS

INTERCONVERSION OF MASS AND ENERGY. The possibility of the interconversion of matter and energy was raised by Einstein in 1905 as part of his Theory of Relativity with his now famous formula: $E = mc^2$. E is the amount of energy released, m is the amount of mass converted to energy, and c is the velocity of light. Since c^2 is a very large figure, E becomes very large. Enormous amounts of energy are released from the conversion of small amounts of matter.

ATOMIC FISSION is the process where a large, heavy atom splits spontaneously under impact of neutron bombardment into smaller atoms, with the conversion of some of the original matter to energy. Three materials that will split immediately by impact of thermal (slow) neutrons are: $_{235}^{92}U$, $_{239}^{94}Pu$, and $_{233}^{92}U$.

In the process of atomic fission, a large unstable atom bombarded by one neutron splits into two smaller atoms plus 2 or 3 neutrons, releasing enormous energy. For every one neutron put in, two or three neutrons are given off. If one neutron produces three, three could trigger nine, nine could propogate 27, etc., thus initiating a rapidly increasing chain reaction. However, not every neutron produced will strike a nucleus. After all, the atom is mostly empty space. Enough fissionable material (critical mass) must be brought together to insure that sufficient neutrons will strike nuclei for the reaction to continue.

THE ATOMIC BOMB. In an atomic bomb, two subcritical masses are suddenly blown together to produce a supercritical mass.

NUCLEAR POWER REACTORS have three essential parts: fuel rods, moderators, and control rods.

(Continued on page 68)

ADDITIONAL INFORMATION

To illustrate the enormous amounts of energy released from the total conversion of small amounts of matter, observe the following: 1 g of matter, if converted entirely to energy, is equivalent to the energy released by the explosion of 20,000 tons of TNT. This was the energy of the Hiroshima bomb.

Elements that split immediately by the impact of thermal (slow) neutrons are not common however. Uranium ore is about 99.3% uranium-238, which is not fissionable at ordinary temperatures. Only 0.7% of the fissionable isotope uranium-235 is present. Plutonium-239 is a man-made element, and uranium 233 is even rarer.

To make the atomic fission bomb, uranium-235 and plutonium-239 were used. Uranium 235 was separated from the element uranium 238 at Oak Ridge, Tennessee by using Graham's Law of Diffusion and allowing the slight difference in molecular weight to effect diffusion through semipermeable filters. At Hanford, Washington, uranium 238 was subjected to neutron bombardment and, after two steps of beta decay, resulted in plutonium 239, which may be split by either fast of slow neutrons and needs no moderator.

A typical fission reaction may be:

$$_{235}^{92}U + _{1}^{0}n \rightarrow _{236}^{92}U \rightarrow _{141}^{56}Ba + _{92}^{36}Kr + 3\,_{1}^{0}n + \text{energy}$$

In a *subcritical mass*, not every fission produces another fission, and no reaction occurs. In a *critical mass*, each fission produces an average of one new fission. (This is the level of the peacetime reactor.) In a *supercritical mass*, each fission produces two or three new fissions, and a self-sustaining chain reaction occurs. (This is the level of the atomic bomb.) (The critical masses of the first atomic bombs were roughly about a pound.)

The fuel rods in nuclear power reactors are made of enriched uranium-235 or plutonium-239. The moderators surround the fuel rods, and form a large cell block into which the fuel rods are poked. Their purpose is to slow down the rate of neutron travel. They are usually made of graphite or heavy water. Control rods interfinger between the fuel rods and are poked into the cell block at right angles to the fuel rods. Their purpose is to absorb neutrons. They are usually made of cadmium or boron. When the control rods are completely pushed in, the reactor is shut down. The more the control rods are pulled out, the faster the reaction proceeds.

(Continued on page 68)

EXPLANATIONS

1. A number of fissionable materials exist.

2. Uranium-235 is separated from uranium-238 by a process involving the slight difference in their rates of diffusion.

3. This need not be the case in every uranium-235 fission.

4,5,6. By definition of subcritical, critical, and supercritical mass. Of the three, critical mass is the most common term.

7. Moderators in atomic reactors are usually graphite or heavy water.

8. Controls rods are usually made of cadmium.

9. The atomic bomb could be more accurately termed the fission bomb.

10. In this process, some of the binding energy is released.

11. Plasma consists of a gas-like cloud of electrons and positive ions. It exists at high temperatures.

12. The matter-energy relationship is given in the Einstein equation: $E = mc^2$.

13. Any such self-sustaining reaction may be termed a *chain* reaction.

Answers

fission	1
235	2
3	3
subcritical	4
critical	5
super-critical	6
the moderator	7
control rods	8
atomic bomb	9
atomic fusion	10
plasma	11
20,000	12
chain	13

ATOMIC FUSION is the process where several smaller atoms combine their nuclei to form one larger atom with conversion of the excess matter to energy. For example:

$$_2^1H + _3^1H \rightarrow _4^2He + _1^0n + \text{energy}$$

This process produces even more energy than atomic fission (about $3\frac{1}{2}$ times more energy per pound).

The sun is an atomic fusion furnace, it combines hydrogen to form helium.

THE FOURTH STATE OF MATTER—PLASMA. We are accustomed to the three commonplace states of matter—solid, liquid, and gaseous. These are really expressions of the amount of kinetic energy of matter. Plasma, the fourth and most energetic state, thrives only at elevated temperatures (100,000°F or above) and does not even begin to exist until around 5000°F. Consequently, it is extremely rare on the earth. Plasma consists of hot charged particles: negative electrons and positive ions. It has been estimated that 50% of the matter of the universe exists in the plasma state.

DETECTION AND MEASUREMENT. Three main devices are used to detect radioactive particles: the cloud chamber, the scintillation counter, and the Geiger counter. Two standards of measurement are used: the curie and the roentgen.

Susceptibility of living organisms to damage from radioactive particles varies directly with amount of body area exposed and length of exposure time; but inversely with individual resistance differences. Perhaps it should be pointed out that we don't know as much as we would like to about biologic response to radiation.

A certain amount of energy is released in atomic fusion because the *binding energy* of the newly-formed nucleus (the energy that holds it together) is less than the binding energy of the original particles.

A fusion reaction requires a higher temperature to start out with. Starting at about 30,000°F, the reaction quickly becomes chainlike at 1,000,000°F or more. Thus we set off an atomic fission bomb first, and this provides enough heat to start an atomic fusion bomb (hydrogen bomb).

A great effort is being made to control the violent nature of plasma and provide controlled fusion as an energy source. The primary problem has been what to keep it in. For short periods of time a strong magnetic field (or *magnetic bottle*) has been used, but this technique has proved to be extremely difficult.

In the *cloud chamber*, radioactive particles ionize vapor molecules and leave visible trails. Against a black background, these trails can be photographed.

In the *scintillation counter*, penetration of fluorescent material by radioactive particles produces a flash of light which is recorded at the other end by a photoelectric cell.

In the *Geiger counter*, incoming radioactive particles ionize gas molecules, permitting a momentary flow of current. This current flow is usually registered on a meter. If boron trifluoride (BF_3) is the gas used in the Geiger tube, incoming neutrons will release positively-charged alpha particles, and the Geiger tube may be used to count (neutral) neutrons.

STANDARDS OF MEASUREMENT. One *curie* is the activity of one gram of radium (3.7×10^{10} disintegrations/sec). It is often subdivided into millicuries (one thousandth of a curie) or microcuries (one millionth of a curie).

The *roentgen* indicates amount of energy of the radiations (it is the amount necessary to produce 1 electrostatic unit of ions in 1 cubic centimeter of dry air at STP.)

Radiation dosage is measured in *rads*. One rad produces the dissipation of 100 ergs of energy per gram of organic matter. (On this basis, 1 roentgen dissipates 83 ergs/g.)

18 THE SOLAR SYSTEM I
SELF-TEST

1. An earth-centered solar system was envisioned by the Egyptian astronomer _____ .

2. A sun-centered solar system with circular paths was proposed by _____ .

3. The paths of the planets were found to be ellipses by _____ .

4. The Foucault pendulum experiment was evidence of the
   a. rotation of the earth.
   b. revolution of the earth.

5. Stellar parallax is evidence of the
   a. rotation of the earth.
   b. revolution of the earth.

6. Stellar aberration is evidence of the
   a. rotation of the earth.
   b. revolution of the earth.

7. The deflection of fluid movement from a N-S line due to the rotation of the earth is
   a. stellar parallax
   b. the Coriolis effect
   c. stellar aberration

8. Apparent movement of a nearby object against a background of distant objects, through motion of the observer, is
   a. stellar parallax
   b. the Coriolis effect
   c. stellar aberration.

9. The times when the sun is farthest north or south are called _____ .

10. The times when the sun crosses the equator are called _____ .

11. The point where the earth is farthest from the sun is the earth's _____ .

12. The point where the earth is closest to the sun is the earth's _____ .

13. At the new moon phase, the moon is _____ illuminated.

14. At the dark crescent (gibbous) phase, the moon is _____ illuminated.

15. At the full moon phase the moon is _____ the sun.

16. Spring tides are unusually _____ tidal ranges.

17. Each synodic month, the moon returns to the same position with respect to the _____ .

18. The earth passes through the shadow of the _____ during a solar eclipse.

19. Seasons are due to the _____ of a planet's axis.

1. _____
2. _____
3. _____
4. _____
5. _____
6. _____
7. _____
8. _____
9. _____
10. _____
11. _____
12. _____
13. _____
14. _____
15. _____
16. _____
17. _____
18. _____
19. _____

70 The Solar System I

BASIC FACTS

EARTH, MOON AND SUN.
Radius of earth = 3,950 mi
Diameter of earth = 7,900 mi
Circumference of earth = 24,000 mi
Distance to moon = 238,860 mi
Distance to sun = 92,956,000 mi
Diameter of moon = 2,160 mi
Diameter of sun = 864,000 mi

Ptolemy thought the earth to be the center of the solar system, and that the paths of the heavenly bodies were large circles (cycles) with small loops (epicycles).
Copernicus thought the sun to be the center of the solar system, and that the paths of the planets were circles.
Kepler also considered the sun the center of the solar system, but that the paths of the planets were ellipses.

The following are evidences of the earth's rotation (movement about an interior axis):
1. Foucault pendulum experiment.
2. Coriolis effect.
3. Rotation of weather eddies.
4. Shape of the earth.
5. Precession of the earth's axis.

The following are evidences of earth's revolution:
1. Stellar parallax
2. Stellar aberration

ROTATION, REVOLUTION, AND THE CALENDAR.
1. A complete rotation of the earth equals one *day* (sidereal day, solar day).
2. A complete revolution of the moon equals one *month* (sidereal month, synodic months).
3. A complete revolution of the earth equals one *year* (sidereal, anomalistic, tropical, lunar years).

The *solstices* are the times when the sun is farthest north or south (approximately June 21 and December 21). At noon on these days, the sun is directly overhead $23\frac{1}{2}°$ north (or south) of the equator.

(Continued on page 72)

ADDITIONAL INFORMATION

The paths of the planets, if plotted in the sky throughout the year, appear to trace out a large loop with smaller loops imposed on it as shown in the figure. Thus, Ptolemy was merely describing what he saw. Predictions made from this theory were fairly accurate and it was "the valid scientific astronomy hypothesis" for thousands of years. As time went on, however, it simply did not fit the observations. The simpler theory of Copernicus gained in favor and finally Kepler's modification was found to fit the facts precisely.

KEPLER'S THREE LAWS OF PLANETARY MOTION.
1. The paths of the planets are ellipses.
2. The speeds of the planets are inversely proportional to their distances from the sun (faster when closer, slower when farther).
3. The distances (cubed) of the planets are proportional to their periods (squared). Thus, observation of a planet's period of revolution enables its distance from the sun to be calculated.

By Kepler's Laws, the paths, speeds, and distances of the planets became known.

EVIDENCES OF EARTH'S ROTATION. *Foucault* hung a heavy weight from a very long wire and set it to swinging. Its path of swing was straight, but its tracings on the floor changed with time as the earth turned beneath it. The *Coriolis effect* is the deflection of fluid movement from a direct north-south line, due to the rotation of the earth. Deflection is to the right in the northern hemisphere and to the left in the southern. *Weather eddy* rotations (cyclones rotate clockwise and anticyclones counter-clockwise in the Northern Hemisphere) are reversed in the Southern Hemisphere. The *shape of the earth* exhibits polar flattening and equatorial bulging due to its rotation. The direction of the earth's axis "traces" a circle in the sky (with time), showing *precession*. Polaris was not the pole star in the past, nor will it be in the future.

EVIDENCES OF EARTH'S REVOLUTION. *Stellar parallax* is an "apparent movement" of a nearby star, against a background of more distant stars, due to the motion of the observer (not the nearby star). *Stellar aberration* is an "apparent movement" of a star in the opposite direction of the observer's motion due to the motion of the observer (not the star). Umbrella analogy—standing still in rain coming straight down, you hold an umbrella directly overhead; if you start to walk forward, the rain appears to come in at you, but it is you who moved.

(Continued on page 72)

EXPLANATIONS

1. The earth-centered solar system of Ptolemy was used for over 1000 years.

2. Copernicus revived this idea held by some of the ancient Greeks. Its modern acceptance stems from him.

3. Kepler tried to mathematically prove a circular sun-centered solar system, but finally abandoned it for ellipses.

4. The Foucault pendulum with time lines gradually moving under its swing evidenced rotation of the earth.

5. Stellar parallax (apparent movement of a star when seen from aphelion and perihelion) evidences revolution of the earth.

6. Stellar aberration (the umbrella analogy) evidences revolution of the earth.

7. The fluid lines would not be deflected as they are in the Coriolis effect if it were not for the rotation of the earth.

8. See the explanation to question 5.

9. At summer solstice, the sun is directly overhead $23\frac{1}{2}°$ North of the equator; at winter solstice, $23\frac{1}{2}°$ South.

10. Day and night are everywhere equal in length at the equinoxes.

11. All planets have aphelions.

12. All planets have perihelions.

13. At new moon, the moon is in the earth's shadow and is dark and usually not seen.

14. At dark crescent, the moon is one-quarter dark and three-quarters illuminated.

15. The full moon gives total reflection because it is opposite the sun.

16. At spring tides the high tides are very high and the ranges between low and high tides are unusually high.

17. The term synodic comes from the Greek *synodos* meaning a meeting. The synodic month refers to a meeting (conjunction) of sun and moon.

18. The earth passes through the shadow of the moon, and the moon obscures our view of the sun, during a solar eclipse.

19. Seasons are due to the tilt of a planet's axis. If a planet's axis has no tilt (from 90° to its ecliptic) it has no seasons.

Answers

Ptolemy	1
Copernicus	2
Kepler	3
a	4
b	5
b	6
b	7
a	8
solstices	9
equinoxes	10
aphelion	11
perihelion	12
not	13
$\frac{3}{4}$	14
opposite	15
high	16
sun	17
moon	18
tilt	19

Equinox—Times when the sun crosses the equator. (March 21 and Sept. 23)

PHASES OF THE MOON. The *new moon* (dark phase) occurs when the moon is towards the sun (relative to the earth).

The *light crescent* occurs when one-quarter of the moon is illuminated.

Quarter is when one-half the moon is illuminated (sun and moon are 90° apart, relative to the earth).

Dark crescent occurs when the moon is three-quarters illuminated (gibbous phase).

Full moon (light phase) occurs when the moon is opposite the sun (relative to the earth).

TIDES. The cause of tides is the gravitational attractions of the moon and sun. The moon is closer and causes 70% of the force of all tides. The force is greatest at the ocean surface nearest the moon, less at the center of the earth, and least at the ocean surface on the earth's side opposite the moon. The ocean is pulled away from the earth on one side and the earth pulled away from the ocean on the other. This produces two high tides which (by rotation) are about $11\frac{1}{2}$ hours apart. Low tides occur inbetween.

The vertical difference between successive high and low tides is the tidal range.

Spring tides have unusually large tidal ranges. They occur at new and full moons when sun and moon pull together (syzygy).

Neap tides have unusually small tidal ranges. They occur at 1st and 3rd quarters when sun and moon oppose at 90° (quadrature).

Tidal lag is due to the inability of the oceans to respond as rapidly as rotation occurs.

ROTATION, REVOLUTION, AND THE CALENDAR. The difference between the lengths of time for the various kinds of days, months, and years depends on the point of reference. A *sidereal day* is the time it takes a star to return to the same position in the sky (24 hours). A *solar day* is the time it takes the sun to return to the same position in the sky (23 h 56 min 4.09 sec). A *sidereal month* is the time it takes the moon to return to the same position with respect to the stars (27.3 days). In a *synodic month*, the moon returns to the same position with respect to the sun (29.5 days). It is also the time between consecutive new moons. In a *sidereal year*, the earth returns to the same orbital position with respect to the stars (365 days 6 h 9 min 10 sec). In an *anomalistic year*, the earth returns to the same orbital position with respect to the sun—time between repetition of perihelion or aphelion (4 min, 47 sec longer than sidereal year). A *tropical year* is the time it takes for the sun's inclination (at noon) to return to the same degree—time between repetition of solstices or equinoxes (20 min 36 sec shorter than sidereal year). A *lunar year* is the time it takes for the appearance of twelve new moons (11 days shorter than sidereal year).

The phases of the moon and the rise and fall of tides follow the synodic month. The *time zones* of the earth arise from the fact that the earth rotates a full 360° in 24 hours. Thus every 15° of longitude gives one hour time difference. Going west (with the sun) one gains time. Going east (against the sun) one loses time. At the *International Date Line* (180° or 12 hours around the world) one gains or loses a day.

Eclipses occur only infrequently, when the earth, moon, and sun are in a straight line. They may be total or partial, depending on the moon's distance from the earth. The inner cone of shadow producing totality is called the *umbra*. The outer cone producing partial shadow is called the *penumbra*. Lunar eclipses last longer (hours instead of minutes) and are visible over a wider area (half the earth instead of 160 miles for a total eclipse and 2,000 miles for a partial eclipse). In an *annular eclipse* of the sun, the moon is too far from the earth to completely block the view, and a ring of the sun is seen around the moon.

Seasons are due to the tilt of a planet's axis away from a right angle (the normal) to its plane of revolution. In winter the sun is low on the horizon. The rays spread out over a larger area, and it is colder (despite the fact that during winter in the Northern Hemisphere the earth is passing its perihelion and is closer to the sun than in the summer). The earth's tilt is $23\frac{1}{2}°$. A planet that has no tilt has no seasons.

19 THE SOLAR SYSTEM II

SELF-TEST

1. In size, the inner planets are _____.

2. In mass, the outer planets are _____.

3. In density, the outer planets are _____.

4. The planet, between the earth and the sun, with a dense cloud cover is _____.

5. The planet with rings is _____.

6. The planet just beyond the earth (from the sun) is _____.

7. The planet discovered by accident was _____.

8. The "great red spot" is found on the planet _____.

9. The outermost planet is _____.

10. The distance from the sun of the planets' orbits is (approximately) predicted by
    a. Bode's Law
    b. Roche's Limit

11. A satellite will break up if closer to a planet than 2.44 times the planet's radius. This is
    a. Bode's Law
    b. Roche's Limit

12. Occupying the space between Mars and Jupiter are the _____.

13. If the object is directly between the earth and the sun, the conjunction is described as _____.

14. If the object and the earth are 90° apart as seen from the sun, the object is at _____.

15. *Syzygy* occurs at
    a. superior conjunction
    b. inferior conjunction
    c. opposition

16. The surface of the sun that we see is the _____.

17. The layer of the sun producing its color effects is the _____.

18. The atmosphere on the moon is _____.

19. An essential difference between the molten state hypothesis of the origin of the solar system and the dust cloud hypothesis is that in the former the planets were formed from the sun, while in the later the planets and the sun were formed _____.

1 _____
2 _____
3 _____
4 _____
5 _____
6 _____
7 _____
8 _____
9 _____
10 _____
11 _____
12 _____
13 _____
14 _____
15 _____
16 _____
17 _____
18 _____
19 _____

BASIC FACTS

BASIC PLANETARY FEATURES.
1. Nearly all the planes of the planetary orbits lie within $3\frac{1}{2}°$ of the plane of the earth's orbit around the sun (the ecliptic).
2. Nearly all of the motions of the solar system are counter-clockwise. The few exceptions probably indicate disturbance or capture.
3. The planets are considered to possess 98% of the angular momentum of the solar system.
4. Two groups of similar planets exist:

The *inner planets* are smaller in size, lower in mass, higher in density, faster in revolution, slower in rotation, and have fewer moons than the outer planets. Mars and Mercury, the smallest, have no cloud cover and consequently no oceans since their gravity is too low to retain an atmosphere. This applies to satellites also. Their normally low size and mass do not usually retain an atmosphere.

The *outer planets* have larger masses which create dense, turbulent, cloud covers (thousands of miles thick) composed of methane and ammonia. This accounts for their low densities; only a small central core is solid. Their rapid rotations cause polar flattening, equatorial bulging, and banding of the clouds. Temperatures range from $-130°C$ ($-200°F$) on down towards absolute zero ($-273°C$).

SOLAR SYSTEM LOCATION TERMS. The *elongation(e)* is the angle between objects in the sky as seen from the earth.

In an *inferior conjunction*, an object is between the earth and the sun.

$e = 0$; new phase; E–O–S

(Continued on page 76)

ADDITIONAL INFORMATION

INDIVIDUAL PLANETARY FEATURES AND OTHER SOLAR SYSTEM MEMBERS. Two laws apply to planets and satellites. *Bode's law* predicts, by mathematical formula, the position of the planets outward from the sun. *Roche's limit* is the minimum radius that a satellite can have if it is not to break up under gravitational stress. This limit is 2.44 times the planet's radius.

The temperature on the sunward side of *mercury* is around 650° C, (at which tin and lead melt) while the other side approaches absolute zero ($-273°$ C). The plane of its orbit is 7° away from that of the earth.

Venus has dense clouds which reflect 60% of all incident light. Its surface is never seen. Radar indicates a very slow clockwise rotation. Interplanetary probes have shown its surface temperature to be about 427°C.

Mars evidences seasonal changes. Its polar cap expands and shrinks while brown areas turn greenish-gray. Pictures taken by an interplanetary probe show a barren, crater-filled landscape like the moon; no magnetic field was recorded. With a low density, this indicates a smaller, entirely solid iron core.

Jupiter shows no seasons, since its axis is not tilted. A semipermanent feature is the "great red spot" which is over 1,700 miles in diameter. Its outer four moons rotate clockwise.

Saturn's rings are made up of small grain-size particles, not solid sheets. They are inside Roche's Limit, which probably accounts for their existence. The outermost ring rotates clockwise.

Uranus was discovered by accident. The planet is "lying on its side," as its axis is only 8° from its ecliptic.

Neptune was predicted, searched for, and found because of irregularities its gravitational attraction caused in the orbit of Uranus.

Pluto likewise was predicted, searched for, and discovered because of irregularities in the orbit of Neptune.

A tenth planet may exist.

The *asteroids* (planetoids) are the over 1500 planetary fragments (or a mass that never formed a planet) between Mars and Jupiter where Bode's law predicts a planet should be. *Comets* largely consist of frozen gases. Their tails always point away from the sun. *Meteors* (called meteorites when they reach the ground) are solid bodies. Most of them are small. The earth passes through swarms of them several times a year. They may be the result of the explosion that formed the asteroids. Comet and meteor orbits do not match planetary orbits. Comets loop far outside.

(Continued on page 76)

EXPLANATIONS

1. The inner planets are much smaller in size.

2. The outer planets are greater in mass.

3. The outer planets are largely gaseous, and so have lower density.

4. Between earth and sun are Mercury and Venus. Mercury has no atmosphere, and Venus has a dense cloud cover.

5. The rings of Saturn are made up of small particles.

6. Mars has two moons.

7. Uranus was discovered by accident by Sir William Herschel when he was mapping the sky.

8. The "great red spot" is over 1700 miles in diameter.

9. Pluto was discovered by Clyde Tombaugh in 1930.

10.
Planet	M	V	E	M	Ast.
Bode's distance	37.2	65.1	93	130.8	260.4
Known distance	37	67	93	142	270

	J	S	U	N	P
	483	930	1804	—	3,608
	483	886	1780	2,790	3,670

11. Roche's Limit probably accounts for the rings of Saturn.

12. The asteroids (or planetoids) are between Mars and Jupiter, located where Bode's law would predict a planet.

13. If it is located beyond the sun it is in superior conjunction, if it is this side of the sun, it is in inferior conjunction.

14. A quadrant is 90° of a circle, hence the term.

15. Syzygy occurs at all three. The only planetary position it does not occur at is quadrature.

16. The term photosphere means light sphere.

17. The term chromosphere means color sphere.

18. The atmosphere on the moon is non-existent, or very nearly so.

19. In the dust cloud hypothesis, the planets and the sun formed together (at the same time) from the same condensing gas and dust clouds. In the molten state hypothesis, the sun is older and the planets formed from it.

Answers

smaller	1
greater	2
lower	3
Venus	4
Saturn	5
Mars	6
Uranus	7
Jupiter	8
Pluto	9
a	10
b	11
asteroids	12
inferior	13
quadrature	14
all three	15
photosphere	16
chromosphere	17
nonexistent	18
together	19

	Distance to Sun (in millions of miles)	Diameter (in miles)	Relative Mass (earth = 1)	Density	Revolution (sidereal)	Rotation	Number of Moons
Mercury	37	3100	0.05	5.1	88 d	60 d	0
Venus	67	7600	0.81	5.0	225 d	243 d	0
Earth	93	7900	1.00	5.5	1 yr	23h 56min	1
Mars	142	4200	0.11	4.0	1.88 yr	24h 37min	2
Jupiter	483	88,000	318.0	1.3	12 yr	9h 50min	12
Saturn	886	75,000	95.0	0.7	29.5 yr	10h 14min	9
Uranus	1,780	31,000	15.0	1.7	84 yr	10h 45min	5
Neptune	2,790	28,000	17.0	2.0	165 yr	15h	2
Pluto	3,670	4000(?)	?	?	248 yr	5 d (?)	0(?)
Sun		864,000	332,000.0	1.4		25-35 d	
Moon	93	2160	0.01	3.39	27.3 d	27.3 d	

In a *superior conjunction*, an object is beyond the sun.

$e = 0$; full phase; E–S–O

Quadrature occurs when the object and the earth are 90° apart as seen from sun.

$e = 90°$; quarter phase; E–S–O

Opposition occurs when the object is on the side of earth opposite from the sun.

$e = 180°$; full phase of moon; S–E–O

When three objects are in a straight line (as in conjunction and opposition), they are at *syzygy*. Conjunctions and oppositions produce the same tidal effects.

THE SUN has 99.8% of the mass of the solar system. It converts 4,000,000 tons of matter to energy each second. It radiates 7800 HP/sq ft/sec, and it has been doing this for about 5 billion years. The sun's four layers are: the interior, the photosphere, the chromosphere, and the corona.

THE MOON always keeps the same side toward the earth. Its low gravity does not retain an atmosphere, and this produces different physical effects than here on earth. Mountains on the moon are higher. The "seas" are lava flows. Most craters have been caused by meteor impact. The largest crater, Clavius, is 146 miles across.

HYPOTHESES FOR THE ORIGIN OF THE SOLAR SYSTEM.
A. Single Star Theories:
   1. Molten State Hypothesis.
   2. Cold Dust Cloud Hypothesis.
B. Double Star Theories:
   1. Near-collision Hypothesis.
   2. Binary Star Hypothesis.

The sun's *interior* produces energy by atomic fusion. Temperatures reach over 15 million degrees C, and pressures reach one billion atmospheres.

The *photosphere* is the surface that we see. Negative ions here absorb light, preventing our seeing any farther in. This absorption reverses the type of spectrum from the continuous emission of the interior to a dark-line, discontinuous, absorption spectrum. Temperatures drop to 6000° C at the top of the photosphere.

The *chromosphere* produces the reddish-orange color (*chromo* means color) from turbulent hydrogen gases. The sonic (sound) waves produced from the turbulence and atomic reaction heat up the gases to around 25,000° C at the top, 3000 miles out from the photosphere.

The *corona* is the glow of light that has been clearly seen out as far as 7,000,000 miles and may extend out beyond the earth. It is best seen during a solar eclipse.

Surface features include sunspots and flares. *Sunspots* are eruptions of magnetic materials (and force lines) that "boiled" to the surface. They produce *solar flares* and prominences that are hurled hundreds of thousands of miles into the corona. Sunspots may disrupt communications and the use of the compass here on earth and cause the auroras (northern and southern lights).

THE MOON rotates in the same length of time that it revolves about the earth, so it always presents the same face to us. The back side of the moon is not visable from the earth. A changing face is presented to the sun, however, and a point on the moon's surface will be in daylight about 15 days and then in darkness about 15 days. Lunar temperatures range from 110° C in the sun to −170° C in the dark areas.

The moon is gradually moving away from the earth. The moon's gravitational force is slowing, and in billions of years, will eventually stop, the rotation of the earth. Both bodies will then keep the same face towards each other. At the same time, the earth is slowing the moon's rotation and revolution. Eventually, the day and month will become equal at around 50 days. The moon may then draw closer to the earth, exceed Roche's Limit, and break up into an orbiting ring of particles.

The *molten state hypothesis* supposes that the planets formed from the sun, after it reached a molten state, by the shedding of rings or shells. The *cold dust cloud hypothesis* supposes that the planets and sun formed together from contracting "dark globules" of a nebula. The sun then became molten by fusion, sweeping away excess gas and dust. The *near-collision hypothesis* supposes that matter was drawn from the sun to form the planets during a near collision with another star. The *binary star hypothesis* supposes that a companion star to sun exploded to form planets.

20 STARS, GALAXIES, AND NEBULAE

SELF-TEST

1. The celestial location system whose coordinates vary with latitude and time is the
   a. celestial sphere system
   b. horizon system

2. Measuring the position of a star (in degrees) N or S of the celestial equator is determining its
   a. declination
   b. right ascension
   c. sidereal hour angle

3. Measuring the angle (in degrees) upward from the horizon to an object in the sky is determining its
   a. altitude
   b. azimuth
   c. sidereal hour angle

4. The distance at which a star has a parallax of one second of arc is a
   a. light year
   b. parsec
   c. absolute brightness

5. The amount of light a star radiates out in all directions is its
   a. luminosity
   b. magnitude
   c. intrinsic brightness

6. Vector combination of Doppler shift movement with across-line-of-sight movement indicates a stars
   a. stellar parallax
   b. proper motion
   c. spectral line intensity

7. Apparent brightness = $\dfrac{\text{intrinsic brightness}}{\text{distance}}$ is an equation for finding
   a. stellar parallax
   b. stellar distance
   c. proper motion

8. Novae and supernovae are
   a. binary stars
   b. exploding stars
   c. pulsating stars

9. Gas and dust clouds (with a few stars) are called
   a. nebulae
   b. galaxies

10. Large star collections (with a few gas and dust clouds) are
    a. nebulae
    b. galaxies

11. Two binary stars whose light intensity varies as they pass in front of one another are called
    a. visual binary stars
    b. eclipsing binary stars
    c. spectroscopic binary stars

12. The most highly developed form of galaxy is the
    a. spiral galaxy
    b. elliptical galaxy
    c. irregular galaxy

13. The galaxy that we are a part of is called
    a. the solar galaxy
    b. Andromeda
    c. the Milky Way

14. Cooler, fainter, older, slower, more numerous (98%) stars belong to
    a. Population I stars
    b. Population II stars

1 _____
2 _____
3 _____
4 _____
5 _____
6 _____
7 _____
8 _____
9 _____
10 _____
11 _____
12 _____
13 _____
14 _____

Stars, Galaxies, and Nebulae

BASIC FACTS

Two methods are used for the *location of celestial objects*: The *celestial sphere system* measures declination and right ascension or sidereal hour angle. The *horizon system* measures altitude and azimuth. Astronomers usually use the celestial system because its coordinates are the same no matter where one is located on earth or when the object is observed. The horizon coordinates of altitude and azimuth vary with latitude and time.

THE STELLAR BRILLIANCE TERMS are:
Brightness—apparent and intrinsic.
Magnitude—apparent and absolute.
Magnitude expresses brightness ratios on a numerical scale.

DISTANCE UNITS. A *light year* is the distance light travels in a year (at the speed of 186,000 mi/sec). This is about 6 trillion miles. A *parsec* is the distance at which a star must be to have a parallax of one second of arc. This is about 3.26 light years. It is an angular distance unit.

STELLAR DISTANCE EQUATION.

$$\text{apparent brightness} = \frac{\text{intrinsic brightness}}{\text{distance}}$$

Apparent brightness is what we see. If we can find intrinsic brightness, we can figure the distance.

Distance may be measured by stellar parallax, in terms of spectral line intensity and intrinsic brightness, in terms of cephid variables and intrinsic brightness, and by doppler shift.

The Doppler shift gives the rate of motion directly toward or away from us. Motion across the line of sight must be determined by visual observa-

(Continued on page 80)

ADDITIONAL INFORMATION

The *celestial sphere system* represents projections on the inside of a sphere with the earth at its center. (Imagine that you could crawl inside a ball and look out.) The stars have fixed positions on this celestial sphere. The sun, moon, and planets change their positions gradually against the background of the fixed stars.

Declination measures the position of a star (in degrees) N or S of the celestial equator (the celestial north pole being the North Star, Polaris). The declination of the celestial equator is 0° and the poles are 90°. Thus it is analogous to latitude.

Right Ascension measures the position of a star in hours, minutes, and seconds of rotation east from a celestial prime meridian called the hour circle (which passes through the celestial equator at the vernal equinox). A complete circle is 24 hours. It is analogous to longitude.

Sidereal hour angle measures the position of a star in degrees west from the vernal equinox. A complete circle is 360°. It, *too*, is analogous to longitude.

The *horizon system* uses the horizon of the earth (terrestrial). Directly above the observer is the zenith. This system describes the position of a heavenly body at a given time with reference to the position of the observer on earth.

Altitude is the angle in degrees measured upward from the horizon to the heavenly body. It is always measured along a vertical circle that would pass through the zenith. *Azimuth* is the number of degrees east from true geographic north to the point where the vertical circle cuts the horizon. Altitude and azimuth are meaningless unless latitude and time of observation are given.

Apparent brightness is the amount of light a star radiates towards us. *Intrinsic brightness* is the amount a star radiates out in all directions.

Apparent magnitude is the apparent brightness ratios on a numerical scale. Typical examples of apparent magnitude are: the sun, -26.8; brightest star, -1.42; faintest star $+22$.

Absolute magnitude is a star's magnitude as it would be if viewed from a distance of 10 parsecs (32.6 light years). Typical examples are: brightest star -10; faintest star $+19$; the sun, $+4.85$ (barely visible).

Luminosity is the brightness of a star compared to the sun.

Stellar parallax is one-half the angle that a nearby star appears to shift when viewed from opposite ends of the earth's orbit. The angular distance unit used is the parsec (*par*allax per *sec*ond). The distance in parsecs is inversely proportional to the parallax in seconds $\left(d = \frac{1}{p}\right)$. This relationship is useful for stars up to 300 light years distant.

(Continued on page 80)

EXPLANATIONS

1. The horizon system coordinates are meaningless unless latitude and time of observation are given.

2. Declination measures a star's position at right angles to the celestial equator. Right ascension and sidereal hour angle measure its position parallel to the celestial equator.

3. Altitude is measured at right angles to the horizon. Azimuth is measured along or parallel to the horizon.

4. A *parsec* is an angular distance measurement and is usually given in degrees (etc.) of arc. A light year is a linear distance measurement and has its equivalent in miles.

5. Intrinsic brightness is the total amount of light radiated. Apparent brightness is only that amount that comes our way. Magnitude is a numerical scale for brightness ratios.

6. Proper motion is the resultant vector from the other two.

7. Stellar distance is determined by this equation if some method will give us intrinsic brightness. Apparent brightness is what we see, so we always know this.

8. Novae (new stars) and supernovae are exploding stars.

9. The word nebulae means filmy, hazy, interstellar clouds.

10. Galaxies may be elliptical or spherical, spiral, or irregular.

11. Light eclipse causes the periodic dimming effect.

12. The spiral galaxy has arms reaching in a direction opposite to that of its rotation.

13. Our solar system is on the outside edge of one of the arms of the Milky Way, a little over half way out from the center.

14. *Population II* stars are cooler, fainter, older, faster and more common (over 98% of all stars are Population II stars).

Answers	
b	1
a	2
a	3
b	4
c	5
b	6
b	7
b	8
a	9
b	10
b	11
a	12
c	13
b	14

Population I Stars
1. In the arms of the central disk.
2. In galactic clusters.
3. In circular orbits.
4. Slower velocity.
5. In many gas and dust clouds.
6. Matter continually replenished.
7. High metal content.
8. Younger, 1–2 billion years.
9. Many 2nd generation stars.
10. Form 2% of all stars.
11. Hot and bright.
12. Hot blue giants go to cool red dwarfs.

Population II Stars
1. In the nucleus, between the arms and in the halo outside the disk. Also in irregular and ellipsoidal galaxies.
2. In globular clusters.
3. In elliptical orbits.
4. Faster velocity.
5. In areas devoid of gas and dust.
6. Matter not often replenished.
7. Lower metal content.
8. Older, 5–6 billion years.
9. All 1st generation stars.
10. Form 98% of all stars.
11. Cooler and faint.
12. Cool red dwarfs expand cool red giants and end as hot blue dwarfs.

tion. The two motions may be combined by vector addition to give the "proper motion" of the star, usually taken over a year. This may be used to predict future positions or reconstruct past positions.

STELLAR PROPERTIES. Most of the stellar properties are derived (directly or indirectly) from study of the spectrum. Temperature, color, composition, mass, diameter, volume and density are all properties used in determining classes of stars.

CLASSES OF STARS are shown on the Hertzsprung-Russell diagram (page 83).

IRREGULAR STARS. Examples of *variable stars* are binary and multiple stars, pulsating stars (cepheids), and irregular variables. *Exploding stars*, novae and supernovae, flare up rapidly and brilliantly for a week or two then become dim. Supernovae are 10,000 times brighter than novae.

COSMIC GROUPINGS. *Nebulae* are gas and dust clouds (with a few stars scattered within). Luminous nebulae have stars in them, or close enough to them, to cause fluorescence or reflection. Dark nebulae blot out areas behind them. They have no nearby star, although occasional stars may "peek" through.

Galaxies are collections of stars (with a few gas and dust clouds scattered within them). Irregular galaxies have no definite shape and are just forming. Elliptical galaxies have the shape of a football. Spiral galaxies have a central flattened disk with 5 or 6 arms that curve backward from the direction in which it rotates. The disk is pinwheel shaped. Scattered globular clusters surrounding it give the whole galaxy a spherical shape. This is the most highly developed form of galaxy.

Star Populations are divided into Population I and Population II star groups.

Spectral line intensity is proportional to intrinsic brightness. Thus, by measuring the intensity of lines in the spectrum of a star, intrinsic brightness may be determined. This used in the distance equation with the star's recorded apparent brightness gives the distance, and is useful for stars up to 2000 light years distant.

Cephid variables are pulsating stars that alternate in their amount of brightness. They are so named because the first such star was discovered in the constellation Cepheus. The period of their alternation (from one maximum brightness to another) is proportional to their intrinsic brightness.

The Doppler shift not only indicates direction of movement towards (blue shift) or away (red shift) from the observer, but the amount of shift of the spectral lines is taken as an indication of the rate of this shift. Rate is equated with distance: The fastest moving, after a given period of time, having reached a point farther away. Such calculations are useful as far as we can see.

DETERMINATION OF STELLAR PROPERTIES. *Temperature* of the surface of a star is indicated by finding the part of its spectrum that its radiation is most intense in. The (*color*) of a star correlates directly with its surface temperature. The *chemical composition* of a star correlates directly with its surface temperature.

Mass can be figured for double or multiple stars that are close enough to gravitationally effect each other.

Volume may be figured by dividing total radiation (intrinsic brightness) by radiation per unit area (temperature). *Diameter* is a geometric property of volume.

Density is mass per unit volume.

The variation in brightness of *variable stars* may be due to periodic eclipse or to actual periodic pulsation in energy release, diameter, and brightness.

Binary Stars may be *visual*, two stars (usually one small and bright, one large and dim) visible; *eclipsing*, two stars revolving around a common center of gravity and the light intensity varying as they pass in front of one another; or *spectroscopic*, too far away to distinguish as two stars, but the shift of spectral lines indicating periodic motion towards us and away from us.

Pulsating stars (mostly cepheid variables) expand their surface area to get rid of energy, and then contract until energy levels build up again.

The *Milky Way* (in which our sun is one of some 100 billion stars) is a spiral galaxy. The central disk is 100,000 light years across, and 10–15,000 light years thick at the center. Our sun is in one of its arms, 27,000 light years from the center. Our galaxy revolves once every 220 million years, and has made one complete turn since the age of the dinosaurs.

21

THE UNIVERSE

SELF-TEST

1. The sun with a 6000° C surface temperature, a yellow-orange color, and with ionized and neutral metals indicated in its spectrum is a
   a. class O star
   b. class G star
   c. class M star

2. A hot blue giant star in the Hertzsprung-Russell diagram would be in the
   a. upper left corner
   b. upper right corner
   c. lower left corner
   d. lower right corner

3. A cool red giant star in the Hertzsprung-Russell diagram would be in the
   a. upper left corner
   b. upper right corner
   c. lower left corner
   d. lower right corner

4. A hot white dwarf star in the Hertzsprung-Russell diagram would be in the
   a. upper left corner
   b. upper right corner
   c. lower left corner
   d. lower right corner

5. Population II stars expand when the amount of hydrogen used up is
   a. 60%
   b. 100%
   c. 15%

6. A Population II star shrinks when its reactions become
   a. endothermic
   b. exothermic

7. Members of the Local Group include
   a. the galaxy Andromeda
   b. the Milky Way
   c. the Large Magellanic Cloud
   d. all the above

8. The atomic bomb presents a physical basis for the
   a. Big Bang Theory
   b. Steady State Theory
   c. Pulsation Theory

9. The cepheid variable star presents a physical basis for the
   a. Big Bang Theory
   b. Steady State Theory
   c. Pulsation Theory

10. The red shift presents a physical basis for the
    a. Big Bang Theory
    b. Steady State Theory
    c. Pulsation Theory

11. Astronomers doubled their concept of the size and age of the universe when they discovered there were two types of
    a. novae
    b. cepheid variables
    c. nebulae

12. According to Einstein's Theory of Relativity, the speed of light in a vacuum
    a. slows down
    b. speeds up
    c. never varies

13. The General Theory of Relativity deals with
    a. nonuniform motion
    b. uniform motion

14. Quasars were discovered only by the use of
    a. the most powerful light telescopes
    b. radio telescopes

1 _____
2 _____
3 _____
4 _____
5 _____
6 _____
7 _____
8 _____
9 _____
10 _____
11 _____
12 _____
13 _____
14 _____

BASIC FACTS

Population I Stars are replenished by gas and dust clouds and come straight down the main sequence.

Population II Stars originate on the lower half of the main sequence, moving towards the center, as they age until 12–15% of their hydrogen is used up. They then expand to cool, red giant stars. When 60% of their hydrogen is used up, they shrink. Crossing below the main sequence, they form pulsating stars or novae and end up as hot blue dwarf stars. Note that cool red dwarf stars may be the end result of the Population I series or the starting point of the Population II series.

The Local Group of galaxies is composed of 17 members and occupies a sphere of space 3 million light years across. It includes Andromeda (or M31—the largest galaxy known), the Milky Way (the next largest), the large and small Magellanic clouds, and 13 other members (most of them known by number).

THEORIES OF ORIGIN OF THE UNIVERSE. *The Big Bang Theory.* Almost all stars and galaxies exhibit a red shift and seem to be moving away from each other and outward from a central point (like dots on a balloon being blown up). It follows that all matter, at one time, must have been at a central point.

The Steady State Theory. Matter is assumed to be constantly forming from energy in space and (as the stars "burn out") energy is being formed from matter. The total amount of mass-energy in the universe remains constant.

The Pulsation Theory. The universe is assumed to alternately expand and contract, like some variable stars. In the expansion phase, more matter is converted to energy. Grav-

(Continued on page 84)

ADDITIONAL INFORMATION

Population II Stars are making more complex elements by the time 12–15% of their hydrogen has been used. This produces more energy, and the star expands to have a greater surface area so that more energy can be radiated off. By the time 60% of the hydrogen is used up, the star is making iron. Up to now, the reactions have been exothermic—giving off heat. Now the reactions become endothermic. The more complicated reactions produce more heat so the star becomes hotter. At the same time, more of this heat is used internally rather than being radiated off, so the star shrinks and moves downward on the Hertzsprung-Russell diagram.

After crossing the main sequence, the energy production becomes variable or erratic and the star is faced with a dilemma like that of the amoeba. The amoeba grows until its bulk reaches a point where its surface area is not large enough to feed it. It then splits into two separate amoebae, each with an adequate volume-surface ratio. The star may solve its excess problem in two ways. It may expand, increasing its surface area and radiating more energy. When the excess energy is "drained off," the star will contract again. These alternating expansions and contractions cause the variable pulsating stars. Or, if the energy release is so rapid that expansion can't keep up with it, the star may explode—shedding excess energy and much matter. The end result is a hot blue dwarf star.

Class	Temperature	Color	Composition
O	80,000°K	blue	hydrogen & helium
B	20,000°K	bluish-white	hydrogen & helium
A	10,000°K	white	hydrogen & ionized metals
F	7000°K	yellow	ionized & neutral metals
G	6000°K	yellow-orange	ionized & neutral metals (sun)
K	5000°K	orange	neutral metals & simple compounds
M	3000°K	red	neutral metals & simple compounds
N,R,S	2000°K	very red	simple compounds
I	1000°K	infrared	simple compounds

The sun is a class G star on the main sequence (6 billion years old and 6% of its hydrogen used) and will follow this pattern, with expansion predicted in about 6 billion years.

THE PHYSICAL BASIS OF THE THEORIES OF ORIGIN. The Big Bang Theory has its physical basis in the predominant red shift. All stars, except those of the constellation Hercules (which the sun is moving toward), show this.

(Continued on page 84)

EXPLANATIONS

1. The sun is a class G star. Plot it on the Population II Hertzsprung-Russell diagram.

2. Hot and blue would be to the left; giant would be upper.

3. Cool and red would be to the right; giant would be upper.

4. Hot and white would be to the left; dwarf would be lower.

5. Expansion occurs when 15% of the star's hydrogen is used up and the star starts making more complicated atoms. These more complex reactions produce more energy.

6. The reaction becomes endothermic (heat-absorbing) when iron begins to be made and 60% of the star's hydrogen is used up. This absorption means that a smaller percentage of the total energy production is radiated and the star shrinks, despite an increasing total amount of energy produced.

7. All four of these plus 13 other members make up the Local Group, which is 3 million light years across.

8. The atomic bomb emphasized the equivalence of mass and energy, and that they can be converted one into the other. The Steady State Theory holds that energy is steadily converted to matter and matter to energy in the universe. Thus it has more direct relation than the Big Bang Theory, despite the name. (Sorry about that!)

9. The cepheid variable stars pulsate and thus form a physical basis for the Pulsation Theory.

10. The red shift, which indicates that all stars are moving away from each other (like dots on a balloon being blown up), is a physical basis for the Big Bang Theory.

11. Two different types of cepheid variables were discovered by Walter Baade. It was because of this discovery that the distance and age scales were changed and Population I and II classifications were set up.

12. Einstein said the speed of light in a vacuum never varies, regardless of motion of the source or of the observer. This underlies astronomical measurement of distance.

13. Nonuniform motion includes accelerated motion and rotational motion. It was this type that Einstein dealt with in his General Theory of Relativity, which treats gravitational distortion in space-time near large masses.

14. The development of radar (in World War II) led to the development of the large radio telescopes that first discovered quasars. The quasars were then investigated with large optical telescopes.

Answers	
b	1
a	2
b	3
c	4
c	5
a	6
d	7
b	8
c	9
a	10
b	11
c	12
a	13
b	14

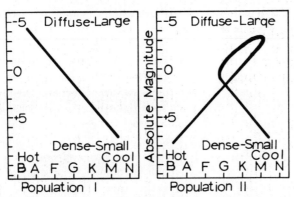

FIG. 21.1. Modified Hertzsprung-Russell Diagram

ity eventually brings about contraction, and more energy converts to matter. The latest estimate is that we are 13 billion years into an 80 billion year expansion cycle.

RELATIVITY, ASTRONOMY, AND SPACE TRAVEL. The Theory of Relativity (see Chapter 15) has several important effects on cosmic research and activity:

1. The constancy of the speed of light in the vacuum of outer space is a basic premise of astronomy today and underlies stellar distance measurements. However, the General Theory of Relativity states that light is slowed down if it passes through the distortion in space-time near a large mass, as it does during eclipses.
2. The shortening of length in the direction of motion, increasingly noticeable at speeds approaching the speed of light, could have a serious effect on equipment where dimensions are critical. Extra-dimensional length may have to be added in compensation.
3. The shortening of time for the rocket traveler may be the "means" of completing long journeys. The traveler, however, who ages only 20 years on a trip at $0.1c$ would return to find that 200 years have passed at his home planet.

One interpretation of a constant radio noise (from all directions) that is heard in the newest, most sensitive antennae is that this may be a portion of the tremendous energy release of the "initial explosion." It may be "tired light" shifted into the radio spectrum.

The Steady State Theory has its physical basis in atomic energy experiments which in recent years have shown the equivalence of mass and energy.

The Pulsation Theory has as its physical basis the pulsating variable stars and makes use of some of Einstein's mathematical concepts.

Geologists, by using radioactive mineral decay, set the age of the earth at 5 billion years (See Chapter 26). The astronomers disagreed. Andromeda was 1 million light years distant, the Mt. Palomar 200 inch telescope was supposed to have a range of 1 billion light years, and the universe was $2\frac{1}{2}$ billion years old. Thus the earth seemed to be older than the universe. In 1941–2, the astronomer Walter Baade, however, using the 200 inch telescope, found that some stars (including variables) were much brighter than others. He named the brighter ones Population I and the dimmer ones Population II. Refiguring on the basis of the new type of variable a little more than doubled distances and ages. Andromeda was now calculated to be 2.2 million light years distant, and the universe was at least 5 billion years old.

Many objects appearing only as dim stars are actually the most distant objects seen and therefore have to be the brightest objects known. These objects, too big to be stars but too small to be galaxies, are called quasars (quasi-stellar light sources). They are only 10–20% as big as the Milky Way, yet they radiate forty times more visible light than it does, and extravagantly pour out radio, ultraviolet, and infrared waves. One is recorded at 10 billion light years away and moving at 80% of the speed of light. Blue-sources (BSO's), giving ultraviolet radiation (indicating very high temperature) and no radio waves, are estimated to be 500 times as numerous. There also exist sources of extremely regular radio pulses called pulsars.

The quasars seem to favor a plane tilted slightly to the plane of the Milky Way. Possible suggested energy sources for the tremendous amount of quasar energy (in order of plausibility) are:

1. Collisions between matter and antimatter galaxies.
2. Gravitational collapse. If a star were 1 million times more massive than the sun, the gravitational compression might cause such tremendous energy releases. Critics say no stars this size have ever been observed, and besides nothing could "get out" in this case—not even light. There would be a "black hole" in the sky.
3. A chain reaction of supernovae. This, however, would require hundreds and thousands of them while we see only one nova or supernova in a century.

22 SPACE EXPLORATION

SELF-TEST

1. The Russian pioneer of rocket theory was _____ .

2. The Romanian-German pioneer of rocket theory was _____ .

3. The American pioneer of rocket theory was _____ .

4. The first liquid-fuelled rocket flight was in the year _____ .

5. The German Army rocket center of World War II was on the Baltic coast at _____ .

6. The first artificial satellite of the earth was called _____ .

7. The first U. S. Satellite, Explorer I, discovered the _____ .

8. The chief former German rocket scientist, who is now in the U. S., is _____ .

9. Most manned rockets today use _____ propellants.

10. Many missile rockets today use _____ propellants.

11. From Newton's Third Law of Motion, the rocket engine is often called a (an) _____ engine.

12. The number of pounds of thrust divided by the number of pounds of fuel used per second is the _____ of a rocket.

13. The number of pounds of fuel divided by the total weight is the rocket's _____ .

14. A rocket with air-lift surfaces that may take off and land horizontally under pilot guidance is called a (an) _____ .

15. The escape velocity of the earth is _____ mi/sec.

16. A landing vehicle, detaching from a space craft and landing on the moon, would not need to be _____ .

17. To remain in orbit, a rocket's _____ must match the pull of gravity.

18. The first manned space flight occurred in the year _____ .

1 _____
2 _____
3 _____
4 _____
5 _____
6 _____
7 _____
8 _____
9 _____
10 _____
11 _____
12 _____
13 _____
14 _____
15 _____
16 _____
17 _____
18 _____

BASIC FACTS

ROCKET DEVELOPMENT. Three men in particular pioneered the development of rocket theory and practice. These men were the Russian Konstantin Tsiolkovsky, the Romanian Hermann Oberth, and the American Robert Goddard.

In 1926, Goddard launched the first liquid-fueled rocket. It traveled 184 feet. He continued his experiments. By 1941, he had flown a rocket of over 900 pounds.

In 1937, a rocket construction and test center was established at Peenemunde on the Baltic coast by the German Army. It was here, in 1942, that the first successful V-2 rocket flight occurred. It went up 53 miles and used 10 tons of fuel to lift a 1 ton load at 3600 mi/hr. In 1945, the V-2 became operational, being fired against London.

In 1949, the United States launched the Viking, the first large American rocket.

During the 1950's, both the United States and the Soviet Union were developing ICBM (intercontinental ballistic missile) programs.

On *October 4, 1957*, the first artificial satellite of earth, Sputnik I, was launched by the U.S.S.R.

On January 1, 1958, the first U.S. satellite, Explorer 1, was launched. It discovered the Van Allen radiation belts. The first lunar impact was achieved in 1959 by the U.S.S.R.'s Lunik 2.

On *April 12, 1961*, the first man launched into space, Major Yuri Gagarin (1934–1968) of the U.S.S.R., orbited the earth in 108 minutes at a speed of 17,400 mi/hr in his Vostok 1 capsule. The first orbiting of the earth

(Continued on page 88)

ADDITIONAL INFORMATION

FROM THEORY TO PRACTICE. *Tsiolkovsky*, in 1903, when the airplane was just beginning, formulated equations concerning rocket travel and published them (along with discussions of space problems such as food, oxygen supply, and space suits). He wrote about liquid fuels and multistage rockets.

Oberth wrote in 1923 about multistage spacecraft, self-cooling rocket motors, rendezvous and refueling in space, and the use of orbiting satellites as way stations. His writings prompted rocket experiments in Europe and America.

Goddard published a pamphlet outlining rocket theory in 1919. Unlike the other two men, he progressed to the practical by actually building rockets. Solid-fuel rockets had, of course, been used in 13th-century China for military use and fireworks display. Goddard, however, built a different type —the liquid-fueled rocket. P. E. Paulet, a South American engineer, had built a working model of a liquid-fueled rocket motor in 1895. Goddard did not know this and developed his own—with very little funds or encouragement.

The German Army had long been interested in rockets, because these were one of the few weapons not forbidden them by the Versailles Treaty. A test center was set up here under Captain Walter Dornberger. He obtained as an assistant Wernher von Braun, a former student of Oberth's. Later, von Braun became chief scientist at Peenemunde and brought Oberth to work there. In the last days of World War II, von Braun and the top German scientists surrendered to the American Army, while the Russian Army captured the rocket installations, technicians, and production engineers—and the German plans for a much larger rocket booster.

Following the war, von Braun and the top German scientists came to the United States and worked at White Sands, New Mexico and Redstone Arsenal, Alabama.

Liquid fuel rockets have certain advantages over solid fuel rockets, in that they produce more energy per pound of fuel, can be restarted, and are easier to control in flight. Solid fuel rockets are easier to store and use. Hybrid rockets have both solid and liquid stages.

Liquid fuels are classified as monopropellants (like H_2O_2) or bipropellants ($H_2 + O_2$).

(Continued on page 88)

EXPLANATIONS

1. Konstantin Tsiolkovsky published papers on rocket theory in 1903.

2. Herman Oberth published in 1923.

3. Robert Goddard published a pamphlet on rocketry in 1919. He went on to build actual rockets.

4. Goddard launched the first liquid-fuelled rocket flight in 1926.

5. Peenemunde was where the V-2 was developed.

6. Sputnik I was launched on October 4, 1957.

7. The radiation belts found by this satellite were interpreted by Dr. James Van Allen, an American physicist, as charged particles trapped in the earth's magnetic field. The belt was named for him.

8. Dr. Wernher von Braun, a student of Oberth and former scientific director at Peenemunde, developed the U.S. Atlas and Saturn rockets.

9. Liquid propellants are used for most manned rockets today because they are more easily controlled in flight and give greater thrust-per-pound.

10. Solid propellants are used for many missile rockets today because they are simpler, more reliable, require less maintenance and are not as expensive.

11. Newton's Third Law of Motion states: "For each and every action there is an equal and opposite reaction." Gases go rearward; rocket goes forward.

12. Scientific Impulse is a measure of what velocity a rocket can obtain.

13. Mass Ratio is a measure of what altitude a rocket can obtain.

14. A rocket plane would be a great simplification of the present process.

15. Escape velocity from the earth is 7 mi/sec, or 25,000 mi/hr. This is the speed necessary to escape the earth's gravity.

16. No streamlining would be needed for a vehicle landing on an atmosphereless body or remaining in orbit in space.

17. A rocket's centrifugal force must balance the pull of gravity on it if it is to remain in orbit.

18. On April 12, 1961, Yuri Gagarin became the first man to successfully orbit the earth.

Answers

Tsiolkovsky	1
Oberth	2
Goddard	3
1926	4
Peenemunde	5
Sputnik I	6
Van Allen belts	7
Wernher von Braun	8
liquid	9
solid	10
reaction	11
specific impulse	12
mass ratio	13
rocket plane	14
7	15
streamlined	16
centrifugal force	17
1961	18

by a U.S. astronaut, Major John Glenn, took place on February 20, 1962. He made three orbits in the Mercury 6 capsule.

Both the United States and the Soviet Union have continued large-scale programs of manned and unmanned space exploration. Both have sent probes to the vicinity of the moon and the neighboring planets. In 1966, both nations safely landed instrument packages on the moon. In October of 1967, the U.S.S.R. parachuted an instrument package to the surface of Venus.

Other nations have conducted far more limited space programs.

Today's rockets operate by one propulsion system only: chemical fuel. *Chemical fuel rockets* may make use of solid propellants or liquid propellants. Other propulsion systems are being experimented with, and may become useful. Included are: free radical systems, nuclear fission, nuclear fusion, plasma jet, ion drive, photon drive, and a solar sail.

ROCKET REACTION. The rocket is often called a reaction engine since it works by Newton's Third Law of Motion—action and reaction. The expanding hot gases escape from the rear of the rocket. This unbalanced force moves the rocket forward.

Rocket measurement terms include specific impulse (proportional to velocity) and mass ratio (determines maximum altitude).

VELOCITY AND TRAJECTORY. *Escape Velocity* is the initial velocity needed for an object to escape from gravitational attraction and go off into space. On earth it is about 7 mi/sec. *Orbital Velocity* is the velocity needed to place the object into orbit. The *trajectory* for an interplanetary rocket must aim in front of the planet, just as a hunter "leads" a moving target.

Future Rocket Propulsion Systems. *Free radicals*, single atoms of the common gases, do not normally exist in nature. If manufactured, they readily combine with something, with the production of considerable heat energy. *Nuclear fission and fusion* should someday provide heat energy for rockets. The weight required for protective shielding is a major problem here. If a *plasma stream* can be achieved, it would provide great energy. The *solar sail* would not carry fuel. Propulsion would come from radiation pressure on a light, aluminized balloon or sail.

ROCKET MEASUREMENTS.

$$\text{Specific Impulse (Isp)} = \frac{\text{lb thrust}}{\text{lb fuel used/sec}}$$

$$= \text{seconds of thrust}$$

$$\text{Mass Ratio (MR or F/W)} = \frac{\text{lb fuel}}{\text{lb total loaded wt.}}$$

Vehicle types depend upon the mission. An orbiting vehicle (or stage) or a vehicle to land on the surface of the moon (or even Mars) would not need streamlining. The rocket plane which could take off and land horizontally from the surface under pilot control may someday augment or supplant the standard rocket.

Guidance systems are seldom pure inertial (preset gyroscopically before lift-off), but provide for in-flight corrections by radio signal from the ground or by "locking" the system optically to a star or target.

Multiple stages are needed for almost all launchings. The velocities from each stage are additive. If, at a moment when a rocket is parallel to the earth's surface, enough energy is added to reach a velocity where centrifugal force balances the pull of gravity, the rocket will assume an orbiting path. Were it not for atmospheric friction, it would need no further energy thrust to stay in orbit.

23 MAJOR MINERALS AND ROCKS

SELF-TEST

Mark each of the following true or false.

1. Minerals are formed either from solution or by metamorphism.

2. The easiest way to identify a mineral is by its chemical properties. *(appearance)*

3. Quartz is the hardest mineral known. *(false)*

4. Breakage along a smooth plane surface is cleavage.

5. Striations help distinguish between the feldspars. *(true)*

6. Amphiboles generally occur with plagioclase feldspar.

7. Ferromagnesians consist of orthoclase and plagioclase.

8. Quartz effervesces and dissolves in hydrochloric acid.

9. Calcite is chemically inert.

10. The formula for clay is SiO_2. *(false)*

11. Orthoclase is (to the unaided eye) the only feldspar in granite.

12. Andesite is a pink, fine-grained equivalent of granite.

13. Pumice is a rock that floats.

14. Diorite is an extrusive fine-grained igneous rock.

15. Sandstone is a deep-water deposit.

16. Shales are made up of the mineral calcite.

17. The rock limestone is made up of clay minerals.

18. Conglomerates have cemented angular fragments.

19. Contact metamorphism changes a mineral by pressure.

20. A gneiss has broad bands consisting of minerals of similar composition.

1. T
2. F
3. F
4. T
5. T
6. ___
7. ___
8. ___
9. F
10. F
11. ___
12. ___
13. T
14. T
15. ___
16. ___
17. ___
18. ___
19. ___
20. ___

BASIC FACTS

A *mineral* is defined as a naturally-occurring chemical compound or element. Minerals are formed from solutions or by metamorphism.

Minerals may be formed from *groundwater solutions*, either hot or cold (hot springs, salt lakes, cave deposits). They may also be formed from *magma solutions* (magma is melted rock below the ground; lava is melted, or once melted, rock at the surface), either by direct crystallization of magma or by crystallization from charged waters and vapors given off by the magma.

Metamorphism occurs when minerals become changed to more stable forms by high heat, chemically active fluids, and pressure (contact with a lava flow, or deep burial). (*Meta* means change and *morphos* means form.)

A mineral may be *identified* by its appearance (the commoner minerals are identified 80% of the time by appearance alone); by its physical properties (15% of the time; identification will have to be by appearance-related, mass-related, constructional, or miscellaneous physical properties); or by its chemical or optical properties (5% of the time; spot chemical tests or polarized-light microscopes will be used). These percentages refer to the average student. Specialists will tend to use the latter group more.

Seven mineral series make up 80% of the rocks of the earth. These are the feldspar, ferromagnesian, mica, olivene, quartz, carbonate, and clay series. Oxygen comprises one-half of the earth's crust (by weight), and silicon one-quarter. The largest mineral group is the silicates (combinations of elements with silicon and oxygen). The next-largest series is the oxides (combinations with oxygen).

(Continued on page 92)

ADDITIONAL INFORMATION

IDENTIFICATION OF MINERALS.
Physical Properties related to appearance are: color (not very reliable as many minerals have several different colors); luster (appearance in reflected light); streak (appearance in powdered form, as by rubbing across an unglazed white porcelain plate); and striations (fine-spaced parallel lines on basal cleavage faces of some crystals).

Physical properties related to mass are hardness and specific gravity.

$$\text{Specific gravity} = \frac{\text{weight in air}}{\text{wt. in air} - \text{wt. in water}}$$

An object may be scratched by any object higher in number on the hardness scale.

The Hardness Scale

1. talc
2. gypsum
3. calcite
4. fluorite
5. apatite
6. orthoclase
7. quartz
8. topaz
9. corundum
10. diamond

Physical properties related to structure are:

Rupture—cleavage (breakage along a smooth plane surface) and fracture (irregular breakage).

Form—crystalline substances have a definite external shape (that they always occur in), determined by the orderly internal atomic arrangement; amorphous substances (like glass) have no fixed external shape.

Constructional properties are brittleness, malleability, ductility, sectility, and tenaciousness.

Miscellaneous properties are radioactivity, fluorescence, magnetism, exfoliation, swelling, piezoelectricity, thermoelectricity, taste, and odor.

The Major Rock-Forming Minerals

Name	Composition	How Distinguished
A. *Igneous rock minerals*		
1. *Feldspar series*—"field-spar"—primary mineral		
orthoclase	$KAlSi_3O_8$	usually pink, no striation
plagioclase°	$NaAlSi_3O_8$ $(Ca,Ba)Al_2Si_2O_8$	never pink, has striations
2. *Ferromagnesian series*—(iron and magnesium)—secondary mineral.		
amphibole	complex silicates or aluminates	black, occurs with orthoclase
pyroxene	complex silicates containing Ca, Na, Mg, Fe, or Al	green, occurs with plagioclase
3. *Mica series*—good basal cleavage—accessory mineral		
muscovite	complex aluminum silicate	light, separates into thin elastic plates

(Continued on page 92)

EXPLANATIONS

1. Both solution and metamorphism form minerals.
2. Appearance is the easiest way to identify minerals.
3. Diamond is the hardest mineral known. It is 10 on the hardness scale, while quartz is 7.
4. Smooth breakage is cleavage; irregular breakage is fracture.
5. Plagioclase has striations, orthoclase does not.
6. Amphiboles are associated with orthoclase, pyroxenes with a plagioclase feldspar.
7. The ferromagnesian minerals are the pyroxenes and amphiboles.
8. Calcite effervesces and dissolves in hydrochloric acid; quartz does not.
9. Quartz is chemically inert. It will only dissolve in hydrofluoric acid. Calcite dissolves in hydrochloric and other acids.
10. SiO_2 is the formula for quartz. Clays are complex silicate minerals.
11. Orthoclase is the only apparent feldspar in granite (5–10% plagioclase may be present, but will not be seen without a microscope).
12. Rhyolite is the pink fine-grained equivalent of granite. Andesite is the gray fine-grained equivalent of diorite.
13. Pumice is light, porous lava that floats. A laboratory sample floated for $3\frac{1}{2}$ days before becoming water-logged and sinking.
14. Diorite is an *intrusive* coarse-grained igneous rock.
15. Sandstone is a near-shore deposit.
16. Shales are made up of clay minerals which are complex aluminum silicates.
17. The rock limestone is made up of the mineral calcite.
18. Conglomerates have cemented rounded fragments. It is breccia which has cemented angular fragments.
19. Contact metamorphism changes minerals by heat and chemically active fluids. It is dynamic, or regional, metamorphism and deep burial that add the effects of pressure.
20. The broad bands of a gneiss are bands of minerals of similar composition. The lighter-colored, lighter-weight feldspars and quartz form one set of bands. The darker-colored, heavier ferromagnesians form another set, alternating between the bands of the first set.

Answers

T	1
F	2
F	3
T	4
T	5
F	6
F	7
F	8
F	9
F	10
T	11
F	12
T	13
F	14
F	15
F	16
F	17
F	18
F	19
T	20

	acidic (sial)	composition		basic (sima)		
Texture	orthoclase the only feldspar	orthoclase + plagioclase	plagioclase the only feldspar	much Fe-Mg, little feldspar	Fe-Mg only	
Crystalline coarse	GRANITE (salmon color)	monzonite	DIORITE (salt & pepper)	gabbro	pyroxenite type	intrusive
Porphyritic (fine)	RHYODITE (pink)	latite	ANDESITE (gray)	BASALT (black)		
Amorphous cellular	←——————— Lava ———————→					extrusive
	pumice "rock froth" (light gray; floats)			scoria "rock slag" (red; oxidized iron)		
glassy	obsidian (black)					
bedded	volcanic tuff, breccia, or conglomerate					

Classification of Igneous Rocks

Major Minerals and Rocks

A *rock* is a naturally-occurring mass of mineral matter, usually covering an appreciable area. Minerals make up rocks. A rock may be made up of only one mineral or of several.

There are three major kinds of rocks: *Igneous rocks* are rocks that were once molten (melted). *Sedimentary rocks* are rocks composed of compacted, cemented sediment. *Metamorphic rocks* are rocks that have been changed in form and mineral content by high heat, chemically active fluids, and (or) pressure.

TYPES OF IGNEOUS ROCKS. *Intrusive* igneous rocks cool slowly, deep below the surface, and thus have time for large crystals to form in them. These rocks are coarsely crystalline or porphyritic.

Extrusive igneous rocks cool rapidly, near or on the surface, and do not have time for large crystals to form. These rocks are finely crystalline or amorphous (cellular or glassy), or else are in wind-blown bedded deposits.

CLASSIFICATION OF ROCKS. Rocks are classified by composition or by texture, or by plotting one against the other. *Igneous rocks* are classified by plotting composition versus texture. *Sedimentary rocks* are classified first by composition, and secondly by texture. *Metamorphic rocks* are classified first by texture, and then subclassified by composition.

Sedimentary Rocks by Composition

Rock Name	Symbol	Mineral Component	Chemical Composition
sandstone	Ss.	Quartz	SiO_2
shale	Sh.	Clays	complex silicate
limestone	Ls.	Calcite	$CaCO_3$
dolostone	Dol.	Dolomite	$CaMg(CO_3)_2$

	biotite	complex aluminum silicate	dark, thin elastic plates
4.	*olivine*	$(Mg, Fe)_2SiO_4$	olive-green color, and occurrence*

B. *Sedimentary rock minerals*

5.	quartz	SiO_2	H = 7, hexagonal crystals, curving fracture, chemically inert
6.	*Carbonate series*		(rhombohedral crystals)
	calcite	$CaCO_3$	H = 3, effervesces in cold HCl
	dolomite	$CaMg(CO_3)_2$	H = 3 to 4, effervesces in hot HCl
7.	*Clays*	hydrous aluminum silicates	H = 2 to 3, soft, earthy appearance and feel

The classification of the igneous rocks is based on being able to distinguish orthoclase feldspar from plagioclase feldspar. That is why the feldspars are termed primary minerals. The usual salmon pink color of orthoclase and the presence or absence of striations is therefore critical.

If individual minerals can be distinguished, the texture is coarse. If only a general color shows, and the rock is not cellular (full of holes like a sponge) or glassy or bedded, the texture is fine grained. If one mineral forms large crystals in a finer background mass, the rock is a porphyry. In coarse prophyries (granite prophyry, etc.), the large crystals make up over 50% of the rock. Sometimes a basic igneous rock mineral occurs in such a large mass it makes up a whole rock. This is named by adding -ite to the mineral term (for example, pyroxen*ite*). Rocks with much feldspar and quartz (sial = silicon + aluminum) are considered acid rocks. Rocks with much ferromagnesians (sima = silicon + magnesium) are considered basic rocks.

SEDIMENTARY ROCKS. Sandstone and shale are often called clastic (fragmental) sediments, and limestone and dolomite are called chemical and biochemical precipitates. Sedimentary rocks may be classified *by texture* (deposited by invading seas) as: conglomerate (a mixture of different-sized rounded fragments) or breccia (a mixture of different sized angular fragments). The breccia has not undergone much erosion and is closer to being in place.

Metamorphic Rocks may be foliated or nonfoliated. Examples of *foliated*, or banded, rocks (several minerals present) are: gneiss (banded by composition from coarse igneous rocks—granite gneiss, etc.) and schist (metamorphism of shale). Examples of *nonfoliated* rocks (only 1 mineral present) are: quartzite (metamorphism of sandstone—much harder than the sandstone was), marble (metamorphism of limestone), and anthracite coal (metamorphism of bituminous coal).

24 ECONOMIC MINERALS

SELF-TEST

1. The ores of iron and aluminum are concentrated by _____.

2. The ores of copper, lead, and zinc are _____ fillings.

3. The major ore of iron, reddish-brown in color, is _____.

4. The secondary ore of iron, yellowish-brown in color, is _____.

5. The black, magnetic ore of iron is called _____.

6. Iron pyrite is more popularly called _____.

7. The major ore of copper is _____.

8. The major ore of lead is _____.

9. The major ore of zinc is _____.

10. The major ore of aluminum is _____.

11. The mineral name for table salt is _____.

12. A principal use of gypsum is to make _____.

13. Fluorite is very useful to make a flux for extracting _____.

14. Hydrofluoric acid is the only acid that dissolves _____.

15. The major use of corundum is for its _____.

16. The major use of talc is to make _____.

17. The major use of graphite is in _____.

18. Diamond has the same chemical composition as _____.

19. The ruby is the red crystalline form of the common mineral _____.

20. The emerald is the green crystalline form of the common mineral _____.

1 _____
2 _____
3 _____
4 _____
5 _____
6 _____
7 _____
8 _____
9 _____
10 _____
11 _____
12 _____
13 _____
14 _____
15 _____
16 _____
17 _____
18 _____
19 _____
20 _____

Economic Minerals

BASIC FACTS

The economically important minerals may be divided into four groups: metallic industrial minerals, nonmetallic industrial minerals, gem stones, and mineral fuels.

INDUSTRIAL MINERALS. The original mineral (if metallic) is called an *ore*. The useful materials in both cases (metallic and nonmetallic) are called products. Thus, hematite is an ore of iron; iron is a product of hematite.

The primary ores of iron and aluminum are the oxide weathering products of sedimentary rocks in humid regions. They become concentrated into economic deposits by the removal of the more soluble compounds of calcium, sodium, potassium, etc.

The ores of copper, lead, and zinc are found in sulfide vein fillings above a magma. These are cracks and crevices above a magma that become filled with waters and gases charged with minerals in solution. As these cool, minerals crystallize out to form a "vein" of mineral matter.

Quartz and calcite veins are common. At times, the metallic ores fill these also. Heavy, metallic, natural elements such as gold and platinum resist weathering and become concentrated by erosion in stream bottoms as placer deposits.

USEFUL NONMETALLIC MINERALS. *Halite* is more familar to us as table salt which is its principal use. It is also a source of chlorine and sodium. It occurs in horizontal evaporite deposits around salt lakes and squeezed up into domes or plugs. *Gypsum* and *Anhydrite* are other chemical salts occuring as cap rock on top of the salt dome. Plaster of Paris is made from them and used in plastering walls and wallboard.

Fluorite occurs in quartz and calcite veins along with galena, sphalerite, and barite. It is a source of fluorine for making hydrofluoric acid, and a flux for extracting aluminum metal from bauxite.

(Continued on page 96)

ADDITIONAL INFORMATION

Hematite has long been the primary ore of iron in the United States. It is mined in Minnesota and the upper peninsula of Michigan. However, the high-grade supplies of this ore are exhausted, and we are now mining taconite, a low grade ore. We are also importing hematite from Venezuela and Newfoundland. *Limonite* (bog iron ore) forms under swampy conditions. It is our secondary source of iron. It is mined in the Appalachian Mountain region, in the Rocky Mountain region, and in the central mineral region of Texas. *Magnetite* deposits are found in Sweden and Russia. The famous Swedish steel is made from this. Magnetite was long called "lodestone" and was used to make primitive compasses. It is prospected for by airborne magnetometer or dip needle.

Chalcopyrite, a compound of copper and pyrite, is the source of most copper sulfide ores. Two copper carbonates are useful both as ores of copper and sources of pigment. Low-cost jewelry and copper plated domes and statues are green, like *malachite*. The azure blue pigment of the artist is ground-up *azurite*. (Incidentally, rouge and lipstick contain the red hematite.)

Galena is unique among metallic ores; it looks like its product. It occurs in square cubic crystals in the same quartz or calcite vein as *sphalerite* (zinc ore), and the nonmetallic industrial minerals fluorite and barite. The miner calls sphalerite "amber jack" or "black jack."

Bauxite looks like enlarged caviar. The pellets are aluminum oxide in a limestone matrix. Aluminum is the most abundant metallic element in the earth's crust, but bauxite is its only practical ore.

Gold occurs in veins. From these veins it erodes down into the streams. Chemically inert and heavy (specific gravity = 19.3), it accumulates as placer deposits.

Silver rarely occurs as an element. It is commonly a black silver sulfide.

Halite is mined from horizontal beds in New York, Michigan, and Kansas. It is extracted from salt domes along the Gulf Coast. Salt, under pressure, is plastic; and, being lighter than the surrounding rocks, it may be squeezed up into domes or plugs. Here it is mined by shafts, or by hot water solution and extraction in the Frasch process.

Gypsum and *anhydrite* are ground to powder, dried to be sure it is in the waterless anhydrite stage, and sold as plaster of Paris. We take this home, put it on a board, and add water taking it back to the gypsum stage. We put it on a wall,

(Continued on page 96)

Economic Minerals

EXPLANATIONS

1. Weathering removes the more-soluble calcium, sodium, and potassium and leaves greater concentrations of iron and aluminum.

2. Hot, mineral-charged waters moving upward into cracks or crevices above the magma fill them with quartz or calcite upon cooling. In some places, the ores of copper, lead, or zinc crystallize out as vein fillings too.

3. Hematite, the reddish-brown major ore of iron, is mined in Minnesota and Wisconsin.

4. Limonite, the yellow-brown secondary ore of iron, is mined in the Appalachian Mountain trend.

5. Magnetite is mined in Sweden and Russia.

6. Iron pyrite, or "fool's gold," is practically worthless.

7. Most of our copper comes from chalcopyrite (an iron sulfide–copper sulfide vein mineral) or from copper ores derived from chalcopyrite.

8. Galena (a lead sulfide vein mineral) is the major ore of lead.

9. Sphalerite (a zinc sulfide vein mineral) is the major ore of zinc and is mined in the same veins and mines as galena.

10. Bauxite, the only ore of aluminum, is an oxide weathering product occuring as light tan pellets in a limestone matrix.

11. The mineral halite (NaCl) is used for table salt when refined.

12. Plaster of Paris (for making plaster walls, plaster board, etc.) is finely ground gypsum or anhydrite.

13. Fluorite (sodium fluoride) can be converted to cryolite (calcium aluminum fluoride) which is the only known flux for extracting aluminum metal from bauxite.

14. Hydrofluoric acid is the only substance that dissolves glass.

15. Corundum is used for abrasive papers, saws, and wheels because of its hardness.

16. Talc is ground up to make talcum powder.

17. Graphite used to be called "white lead."

18. Diamond and graphite are both pure carbon, but diamond has a crystal system that is stable at higher temperatures.

19. Corundum is a common igneous rock mineral; its clear crystalline forms are the blue sapphire and the red ruby.

20. Beryl is even more common in the igneous rocks than corundum. It is the source of beryllium, an element used in making certain types of steel. It may occur in large crystals. (One was measured at 6 feet wide and 32 feet long.)

Answers

weathering	1
vein	2
hematite	3
limonite	4
magnetite	5
fool's gold	6
chalcopyrite	7
galena	8
sphalerite	9
bauxite	10
halite	11
plaster of Paris	12
aluminum	13
glass	14
hardness	15
talcum powder	16
pencils	17
graphite	18
corundum	19
beryl	20

Nonmetallic Economic Minerals

Name	Composition	Appearance	Hardness	Product
gypsum	$CaSO_4 \cdot 2H_2O$	transparent to white or gray	2	plaster of Paris
anhydrite	$CaSO_4$			
halite	NaCl	transparent to white	2–3	salt
fluorite	CaF_2	usually purple, but transparent	4	hydrofluoric acid, flux for Al
corundum	Al_2O_3	light gray	9	abrasives, gems
talc	hydrous magnesium silicate	white	1	talcum powder
graphite	C	black	1–2	pencil lead, lubricant, electrodes

Corundum is an igneous rock mineral. It is used for abrasive papers and wheels because of its hardness (H) of 9 (10 is the hardest).

Talc is at the opposite end of the hardness scale. Its hardness varies from 1 to $2\frac{1}{2}$. It is a metamorphic rock mineral and is ground up, perfumed, and sold as talcum powder. Called "soapstone" at H = $2\frac{1}{2}$, it is used for small statues and sinktops.

Graphite, another metamorphic mineral with a hardness equivalent to talc, has 3 uses: as a marking substance, in carbon electrodes, and as a solid lubricant.

A few pyroxenes and some related silicate minerals are fibrous. They are pulled apart, woven into paper or cloth, and sold under the name *asbestos*.

ROCKS commonly used for *building stone* include limestone, granite, marble, and sandstone. *Chemical stone* includes limestone (cement, flux for steel, road ballast) and sandstone (a basis for glass).

GEM STONES are the rare, clear, crystalline forms of common minerals with hardnesses of above 7. Thus they are both valuable and durable.

allow it to dry for 2 or 3 days (going back to the anhydrite stage) and we have a plaster wall.

As previously indicated, the extraction of aluminum is very difficult. In the 1870's there was only about a pound of aluminum in the world. Then Charles Hall, a chemistry professor at Oberlin College in Ohio, discovered a way to extract aluminum from bauxite using the mineral cryolite. Although natural cryolite is rare, the common, more plentiful *fluorite*, a calcium fluoride, can be made into artificial cryolite, a sodium aluminum fluoride.

Hydrofluoric acid is the only acid that will dissolve glass. It is used to etch and make frosted light bulbs. It has to be kept in paraffin bottles.

There is no lead in a "lead pencil;" it is all *graphite*. The lead pencil was so named because graphite was called plumbago or white lead. Graphite is one of the few nonmetallic elements that will conduct an electric current. Spaced apart, two carbon electrodes produce a bright carbon-arc spark between them. This carbon arc is used in search lights and some projection lamps.

Sulfur is the residue left after the solution of calcium from the gypsum and anhydrite cap rock. The Frasch process for recovery of sulfur was developed at Sulphur, La., where loose, unconsolidated sands prevented sinking a shaft.

THE METALLIC ORES.

Name	Composition	Appearance	Ore of
hematite	iron oxide	red, reddish-brown	iron
limonite	hydrated iron oxide	yellow-brown (dull)	iron
magnetite	iron oxide	black	iron
pyrite	iron sulfide	pale gold-yellow (metallic)	iron
chalcopyrite	copper iron sulfide	brass yellow (metallic)	copper
malachite	hydrated copper carbonate	green	copper (pigment)
azurite	hydrated copper carbonate	blue	copper (pigment)
galena	lead sulfide	gray (metallic)	lead
sphalerite	zinc sulfide	yellow-brown (resinous)	zinc
bauxite	hydrated aluminum oxide	light brown (round concretions)	aluminum
gold	element	golden yellow	gold
silver	silver sulfide, element	dark metallic	silver

25

MAPS

SELF-TEST

1. The general location of a map is told by its _____.

2. The relationship between distances on the map and in the field is shown by the map's _____.

3. The compass direction of the top of a map is always _____, unless otherwise indicated.

4. 1:62,500 is called a representative _____.

5. On a scale of 1 to 62,500; 1 inch on the map represents approximately _____ in the field.

6. One set of common map reference lines shows latitude and _____.

7. Another set of common map reference lines shows township and _____.

8. Man-made features on a map are shown by the color _____.

9. Water features on a map are shown by the color _____.

10. Elevation features on a map are shown by the color _____.

11. A long solid line with even-spaced cross bars on a map indicates a(an) _____.

12. A township has _____ square miles.

13. The NW section in a township is always numbered _____.

14. The SE section in a township is always numbered _____.

15. One-fourth of a section has _____ acres.

16. Map lines connecting all points of equal elevation are called _____ lines.

17. These lines close together indicate a _____ slope.

18. These lines far apart indicate a _____ slope.

19. A map showing where rock formations outcrop at the surface is called a _____ map.

20. A map with contour lines showing changes in the land surface is called a _____ map.

1. _____
2. _____
3. _____
4. _____
5. _____
6. _____
7. _____
8. _____
9. _____
10. _____
11. _____
12. _____
13. _____
14. _____
15. _____
16. _____
17. _____
18. _____
19. _____
20. _____

BASIC FACTS

A *map* is a graphic representation, on paper, of a portion of the earth's crust.

THE ELEMENTS OF A MAP are:

1. A *title*, telling the general location of the map.

2. A *scale*, giving the ratio between distances on the map and the same distances in the field.

3. *Symbols*, which are different geometric forms or pictographs conventionally used to represent specific map features.

4. A conventional *orientation*. The top of a map is conventionally north. This may be shown by *magnetic declination* arrows, one pointing to geographic north (the pole of rotation) and the other to magnetic north (the compass pole). The magnetic declination in degrees east or west of true north (geographic north) is the angle between these arrows. It is printed at the bottom of the map.

5. *Reference lines*, for orientation and location. Common ones used are: *Latitude and longitude* (latitude lines run E–W, but measure distance N–S; longitude lines run N–S, but measure distances E–W) or *township and range* (township lines run E–W, but measure distances N–S; range lines run N–S, but measure distances E–W). Township and range lines cross, at right angles, to form the Township and Range Grid System. Correlation between the two is established by making the base lines of the grid system parallels of latitude and meridians of longitude.

6. *Contour lines*, for showing the shape and elevation of the land surface. A contour line is a line on a map connecting all points of equal elevation.

(Continued on page 100)

ADDITIONAL INFORMATION

MAP SCALES are commonly expressed in two different ways:

1. As a representative fraction (R.F.). For example: $\frac{1}{62,500}$ or 1:62,500 means that one inch on the map represents 62,500 inches in the field. How is this figure obtained? It is close to the number of inches in a mile (63,360 in.), but is a more even and easily useable figure. The difference is 71.6 ft, which is about 0.015 in on the map (or less than a pencil width). The following table shows the scales used by the U.S. Geological Survey in their National Topographic Map Series.

National Topographic Map Scales

Scale	1 in. represents	Quadrangle size (latitude-longitude)
1 : 20,000	about 1,667 ft	$7\frac{1}{2} \times 7\frac{1}{2}$ min
1 : 24,000	2,000 ft	$7\frac{1}{2} \times 7\frac{1}{2}$ min
1 : 62,500	nearly 1 mile	15 × 15 min
1 : 63,360	1 mile	15 × 20–36 min
1 : 250,000	nearly 4 miles	1° × 2°
1 : 1,000,000	nearly 16 miles	4° × 6°

2. As a bar scale. The length of the bar is drawn graphically to scale. Frequently, both methods will be given at the bottom of a map.

Conventional *color patterns* are used for various types of symbols: *Black* indicates man-made features such as buildings, boundaries (national, state, or local), roads, railroads, power lines, mines, quaries, etc. Major man-made features on recent maps may be outlined in *red*. *Blue* indicates water features such as lakes, swamps, streams, glaciers, etc. *Brown* indicates natural land-surface features such as beach deposits, sand dune areas, etc. Topographic contours lines which show the shape and elevation of the land surface are in brown. *Green* indicates vegetation features such as woodland, scrub, orchard, vineyard, etc.

Some conventional *symbols* used are: *Major roads and boundaries* are solid lines. *Unimproved roads, trails, and minor boundaries* are dashed or dotted lines, or dashes or dots interspersed in a solid line. *Buildings* are shown by rectangles or squares. *Schools, churches, and cemeteries* are shown by a flag or a cross on or within the rectangle or square. *Wells* are small circles. If the circle is filled in solid, it indicates an *oil well*; with radiating lines projecting out from it, it is a *gas well*; with a slanted line through the circle, the well is *dry and abandoned*. A *railroad track* is a line with even spaced cross bars. Multiple tracks are indicated by a double line.

MAP LOCATION BY THE TOWNSHIP AND RANGE GRID SYSTEM. This is often called the General Land Office Grid System. *Townships* are numbered north and south from the

(Continued on page 100)

Maps

EXPLANATIONS

1. The title tells the general location of a map. This is usually the first thing looked for.

2. The map scale relates distance on the map to that in the field.

3. The top of a map is always north, unless otherwise indicated.

4. 1:62,000 is called a representative fraction. It may be written $\frac{1}{62,500}$. It expresses the map scale.

5. One inch on the map represents approximately 1 mile in the field if a scale of 1 to 62,500 is used. Actually there are 63,360 inches in a mile.

6. Latitude and longitude lines are used as reference lines on some land maps and on all oceanic charts.

7. Township and range lines are used as reference lines for land measurement maps, oil well location maps, geologic maps, and topographic maps.

8. Black is used to show man-made features on a map, such as buildings, cities, roads, and political boundaries.

9. Blue is used to show water features on a map, such as rivers, lakes, and seas.

10. Brown is used to show elevation features on a map, usually by contour lines but sometimes by shading.

11. A railroad track is shown by a long solid line with even spaced cross bars. The cross bars represent cross ties. Double track is shown by two parallel long lines.

12. A township is 6 miles wide by 6 miles long and hence has 36 square miles.

13. The NW section of a township is always section number 6.

14. The SE section of a township is always section number 36.

15. A section is one mile square and contains 640 acres. A half-section contains 320 acres, and $\frac{1}{4}$ section contains 160 acres.

16. Contour lines are map lines connecting points of equal elevation. All points on a contour line have the same elevation.

17. A steep slope is indicated by contour lines that are close together.

18. A gentle slope is indicated by contour lines that are far apart.

19. A geologic map shows (by colors or symbols) where rock formations outcrop at the surface.

20. A topographic map has contour lines showing changes in the land surface.

Answers

Answer	#
title	1
scale	2
north	3
fraction	4
1 mile	5
longitude	6
range	7
black	8
blue	9
brown	10
railroad track	11
36	12
6	13
36	14
160	15
contour	16
steep	17
gentle	18
geologic	19
topographic	20

A three-dimensional model such as this should have a vertical scale as well as a horizontal one. This vertical scale is called the *countour interval*. If there is a vertical difference, in the field, of 20 ft between points on adjacent contour lines, we say the contour interval is 20 ft. Small contour intervals (5 or 10 ft) are used for flat land and large contour intervals (50 or 100 ft) are used for mountain regions. The vertical scale of a map or cross section is nearly always an *exaggerated scale*. The basic vertical unit is larger than the basic horizontal unit in order for vertical features to be seen at all.

7. A map may have *area symbols* or *colors*, to show areas of similar features. A *geologic map* is an example of this practice. A different symbol or color covers each area where rock formations outcrop at the surface in such a map. A *topographic map* shows the topography (elevation and relief) of the land surface by graphic methods (contours or shading). *Elevation* is height above sea level. *Relief* is the difference in elevation between the highest and lowest points.

township base line. *Ranges* are numbered east and west from the range base line. Each block thus formed is called a "township" and is 6 miles long on each side. Its area is 36 square miles. In giving its location, abbreviations are used, and townships are designated before ranges. (Thus, T2N, R1E—not R1E, T2N.)

Each Township is then divided into mile-square *sections*. The 36 sections in a township are numbered, starting with the upper right corner, and going across and down through 6 rows. Thus, section 1 is always in the northeast corner, section 6 in the northwest, section 31 in the southwest, and section 36 in the southeast corner. Each section is one mile long on a side and its area is 1 square mile. Section numbers are written before township and range. (Thus, S15, T2N, R1E.)

Sections are subdivided into *halves* and *quarters*. These are given compass designations. Three subdivisions into quarters may be made. The smallest subdivision is written first and the numeral $\frac{1}{4}$ may be left off. Thus we have NW $\frac{1}{4}$, of the SW $\frac{1}{4}$, of the SE $\frac{1}{4}$ of Sect. 15, T2N, R1E. This is usually written NW, SW, SE, S15, T2N, R1E (see Figure 25.1). More precise location than this (if needed) is usually given as so many feet N or S and so many feet E or W from a given boundary. A mile section has 640 acres, $\frac{1}{2}$ = 320 acres, $\frac{1}{4}$ = 160 acres, $\frac{1}{8}$ = 80 acres, $\frac{1}{16}$ = 40 acres, $\frac{1}{32}$ = 20 acres.

RULES FOR READING OR DRAWING CONTOUR LINES.
1. Every point on a contour line has the same elevation.
2. Contour lines never cross or split.
3. Contour lines far apart indicate a gentle slope; lines close together indicate a steep slope.
4. Contour lines repeat on each side of a stream or ridge.
5. Contour lines bend upstream crossing a valley. Contour lines bend downstream crossing a divide.
6. A depression is shown by hatchure marks on a contour line pointing into the depression.

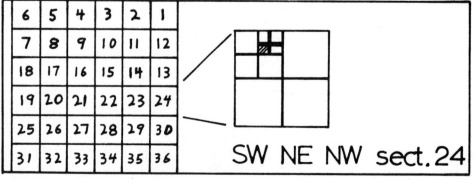

FIG. 25.1. Subdivision of Land Sections.

26 STRATA AND GEOLOGIC TIME

SELF-TEST

1. "The present is the key to the past." This is a statement of the Law of _____, a fundamental law of geology.

2. "All sedimentary beds were originally deposited in a horizontal plane, or nearly so." This is a statement of the Law of Original _____.

3. "In any series of undisturbed beds, the oldest are on the bottom and the youngest on the top." This is a statement of the Law of _____.

4. "Anything that cuts across or intrudes into a series, is younger than that series." This is the Law of _____ Relationships.

5. "Different formations may be correlated by means of the successive fauna and flora found in each." This is the Law of _____ Succession.

6. Sedimentary formations put in order with the oldest beds below and the youngest beds above form a(an) _____ Time Column.

7. If specific ages of beds at different intervals throughout the column can be determined, it can be converted to a(an) _____ Time scale.

8. The most precise and modern method of finding specific dates of formations is by the _____ of minerals.

9. A break in the formation sequence, indicating interruption of sedimentation uplift and erosion, and reburial, marks a buried surface of erosion. This is called a(an) _____.

10. A buried surface of erosion between beds that are parallel above and below is called a(an) _____.

11. A buried surface of erosion between beds that are not parallel above and below (the lower beds being deformed or tilted) is called a(an) _____ unconformity.

12. A buried surface of erosion between different rock types (igneous, sedimentary, metamorphic) is called a(an) _____.

13. The largest divisions of the time column are called _eras_.

14. The second largest divisions of the time column are called _eras_.

15. The third largest divisions of the time column are called _periods_.

16. The time column is divided into the largest divisions on the basis of _____.

17. The largest divisions are subdivided into the second largest divisions on the basis of _____ building.

1 ____
2 ____
3 ____
4 ____
5 ____
6 ____
7 ____
8 ____
9 ____
10 ____
11 ____
12 ____
13 ____
14 ____
15 ____
16 ____
17 ____

BASIC FACTS

The rocks of the earth are formed into an extremely large number of strata or beds. Attempts to put these into some sort of a chronological order have resulted in the building of a relative time column and its conversion into an absolute time scale. These are built upon five basic principles.

FUNDAMENTAL LAWS OF GEOLOGY.
1. Law of Uniformitarianism.
2. Law of Original Horizontality.
3. Law of Superposition.
4. Law of Crosscutting Relationships.
5. Law of Faunal Succession.

These determine the relative position of beds.

RELATIVE TIME COLUMN. Superposition will put the beds of any one outcrop in relative order. Crosscutting will relate the time of faulting or intrusion. Faunal succession will correlate, by fossils, the beds of widely separated outcrops. (An outcrop is where a bed is exposed at the *surface*.) *Correlation* may also result from tracing continuity of outcrop or from comparing lithologic similarity.

ABSOLUTE TIME SCALE. The relative time column gives us relative positions and ages, but we do not know the specific age of any one of the beds. If we can find specific ages, we can convert to an absolute time scale. *Methods used to find precise dates are:*
1. Rates of erosion and sedimentation compared to total sediment thickness.
2. Rate of accumulation of salts in the seas compared to the present total level.
3. Radioactive decay dating of minerals (see Chapter 16).

The first two give only approximations but the latter method is very precise.

SUBDIVISION OF THE TIME SCALE. The column and time scale is too large and unwieldy. Consequently 3 major

(Continued on page 104)

ADDITIONAL INFORMATION

Law of Uniformitarianism. The forces and processes that were active in the past are the same that are active today (with perhaps some changes in magnitude). "The present is the key to the past." We don't need to invoke mysterious unknown forces and processes to explain the geologic effects we see in nature. They were made by forces and processes that are still operating today.

The Law of Original Horizontality. All sedimentary beds were originally deposited in a flat, horizontal plane, or nearly so. Any change from this position indicates disturbance of the beds.

Law of Superposition. In any series of undisturbed beds, the oldest is on the bottom and the youngest on the top.

Law of Cross-Cutting Relationships. Anything that cuts across or intrudes into a series of beds is younger than the series. The beds had to be there before the fault could cut across them or the dike intrude into them.

Law of Faunal Succession. Different formations may be recognized, correlated with beds of similar age, and placed in chronologic order of superposition by the fossil assemblages they contain, or by fossils restricted to certain formations (*index fossils*).

UNCONFORMITIES are interruptions in sedimentation. During the time represented by an unconformity, the land surface is raised above sea level and undergoing erosion. The time represented may be very short or millions of years.

A *disconformity* is a buried surface of erosion between sedimentary formations that are parallel above and below.

An *angular unconformity* is a buried surface of erosion between beds that are not parallel above and below. The lower, older formations were tilted or deformed before erosion.

A *nonconformity* is a buried surface of erosion between different rock types (igneous, sedimentary, metamorphic).

TERMS OF THE TIME SCALE: *azoic* (without life)—no fossils are found in these ancient rocks, *cryptozoic* (hidden life)—many plants and animals (all marine) lived in this time but they formed no shells or hard parts, *paleozoic* (old life), *mesozoic* (middle life), and *cenozoic* (new life).

A *formation* is a mappable rock unit. Several similar formations may make up a *group*. Subdivisions of formations are *members*. Since rock units frequently cut across time boundaries (may be older at one end than the other) they are frequently not equivalent to the other two.

(Continued on page 104)

EXPLANATIONS

1. The Law of Uniformitarianism asserts that the laws and processes that operate today operated uniformly in the past.

2. The Law of Original Horizontality describes the original attitude of sedimentary beds. Even disturbed (tilted, folded, or deformed) beds may be referred back to this original attitude.

3. The Law of Superposition gives the relative order of age of any series of undisturbed beds. It is the basis of stratigraphy (graphing of the strata).

4. The Law of Cross-Cutting Relationships allows us to put in relative time order with a series of beds such things as faults, batholiths, dikes, and sills.

5. The Law of Faunal Succession enables series of beds that are geographically separated (perhaps miles apart) to be correlated and thus worked into the relative time sequence.

6. A Relative Time Column shows the relative order of age of formations but not the precise age of any one.

7. An Absolute Time Scale adds precise dates for a few formations scattered throughout the column.

8. The radioactive decay of minerals (uranium or thorium for the geologist, carbon 14 for the archeologist) gives the most precise modern figures for the dates of rocks.

9. An unconformity is a break or interruption in sedimentation or deposition. It is represented by a buried surface of erosion, indicating an interval when the surface was above sea level, and erosion took place.

10. A disconformity is merely an interruption in a sequence of normal horizontal sedimentation.

11. An angular unconformity frequently (not always) represents a longer time interval than the disconformity. Time must be allowed for the deforming of the lower beds.

12. In a nonconformity, the contact must be one of erosion, not just an original surface contact.

13. Eons are the largest divisions of the time table.

14. Eras are the second largest divisions of the time table.

15. Periods are the third largest divisions of the time table.

16. The absence, presence, and relative abundance of fossils subdivides the time table into Eons.

17. Periods of mountain building, called revolutions or orogenies, subdivide the Eons into Eras. The orogeny is the boundary of the Era.

Answers

uniformitarianism	1
horizontality	2
superposition	3
cross-cutting	4
faunal	5
relative	6
absolute	7
radioactivity	8
unconformity	9
disconformity	10
angular	11
nonconformity	12
Eons	13
Eras	14
Periods	15
fossils	16
mountain	17

subdivisions are made: *Eons* are based on the absence, presence, and relative abundance of living organisms. *Eras* are subdivided by the major orogenies (mountain building episodes). *Periods* are subdivided by the major unconformities (breaks in deposition).

The start of the Paleozoic era is placed at the point where marine invertebrates first secreted calcite shells and thus left hard parts. Life came out of the sea in the Silurian period, when plants and scorpions first invaded the land. The Devonian period was the "age of fish," as the Mesozoic era was the "age of reptiles" (dinosaurs), and the Cenozoic era the "age of mammals" (including man). Each of these "ages" means that this animal group was the most predominant life form at that time, having first developed a little earlier.

The names of the periods are mostly geographic names of the areas of first description. Devonian comes from the Devonshire district of England, mississippian comes from outcrops along the river near St. Louis, and so forth.

Neogene and Paleogene are newer terms proposed to provide a more even time distribution of the Cenozoic (45 M and 25 M as against 69 M and 1 M).

Epochs of the Cenozoic Era

Period		Epoch	Time ago (million years)
Old Terms	New Terms		
Quaternary	Neogene	Present	
		Pleistocene	1
		Pliocene	11
		Miocene	25
Tertiary	Paleogene	Oligocene	40
		Eocene	60
		Paleocene	70

Geologic Column and Time Scale

Era	Period	Time Ago (million years)	Orogeny	Life Forms		
				Animals		Plants
Cenozoic	Quaternary	1	Cascade	Man		hardwood forests
	Tertiary	70	Laramide	mammals		
Mesozoic	Cretaceous			reptiles		conifers
	Jurassic					
	Triassic	225	Appalachian			
Paleozoic	Permian			amphibians		spore-bearing trees
	Pennsylvanian					
	Mississippian					
	Devonian			fish		
	Silurian					
	Ordovician			marine invertebrates	with shell	seaweed
	Cambrian	600	Killarney		no hard parts	algae
Pre-Cambrian		3400				
Formation of Planets		5000				

27 GEOMORPHOLOGY—WATER AND WIND

SELF-TEST

1. The study of land forms of the earth is called _Geomorphology_

2. Lowering the grade, slope, or profile of the land is called _degradation_

3. Weathering is called a passive process because the material is not _transported_

4. Mutual wear between two particles in transit is _attrition_

5. Gouging (bulldozing) is performed by the erosional agent of _glacial ice_

6. The most important agent of erosion is _running H₂O_

7. A stream newly formed on an initial slope has a(an) _____ pattern.

8. A stream that established its course before deformation occurred is _____.

9. In parallel, folded mountain areas, stream drainage patterns are _trellis_.

10. Faults raising resistant formations in the path of a stream cause _____ drainage patterns.

11. A youthful stream carves a(an) _____ shaped valley in cross section. ✓

12. When a stream is in the stage of youth, _____ covers the entire valley floor.

13. Going from youth to maturity, a stream path starts to _meander_.

14. The reason for the effect described in question 13 is because the stream is no longer spending all its energy _____.

15. In a region that is in maturity, the surface is mostly in _slope_.

16. A current that travels parallel to the shore line is called a(an) _longshore_ current.

17. A sand bar that builds across the mouth of a bay is called a(an) _____ bar.

18. A shallow desert lake that is dry or greatly reduced most of the year is called a(an) _____ lake.

19. A crescent-shaped sand dune with the horns pointing downwind is called a(an) _____.

20. Unsorted, unstratified deposits of silt-sized particles, probably wind blown, are called _loess_.

1	
2	
3	
4	
5	
6	
7	
8	
9	
10	
11	
12	
13	
14	
15	
16	
17	
18	
19	
20	

BASIC FACTS

First order land forms (continents and ocean basins) are in part altered into *second order land forms* (mountains, plateaus, and plains) by lithospheric (rock sphere) forces of vulcanism and diastrophism. This will be discussed in Ch. 29. In this chapter, we shall study primarily the effects of atmospheric and hydrospheric forces that produce further alteration to *third order land forms* (valleys, ridges, streams, lakes, and deltas). The study of land forms is called GEOMORPHOLOGY.

THE BRANCHES OF GEOMORPHOLOGY are: *degradation* (the lowering the grade, slope, or profile of the land) and *aggradation* (the raising the grade, slope, or profile of the land by deposition—uplift and mountain building are not included).

THE PROCESSES OF DEGRADATION are: *weathering*, which is the decomposition of materials which are left in place (in situ), a passive process, and *erosion*, which is the decomposition of materials which are transported to another place, an active process.

THERE ARE TWO TYPES OF WEATHERING. *Mechanical* weathering breaks rock into smaller fragments without chemical change. *Chemical* weathering alters the composition of the rocks.

THE AGENTS OF EROSION. (The active mobile agents) include running water (most important), wave action (moves great volumes), glacial ice (extremely powerful), wind (intermittent agent), and gravity (always acting).

HYDROLOGIC CYCLE.

Evaporation → condensation
　　→ precipitation = runoff
　　　　　　+ infiltration

There are two *stream types*: major streams (initial slope—consequent or structure controlled—antecedent and

(Continued on page 108)

ADDITIONAL INFORMATION

TYPES OF WEATHERING (passive)

Mechanical	Chemical
expansion and contraction	solution
exfoliation	suspension
frost wedging (splits rock)	corrosion
plant wedging (splits rock)	

TYPES OF EROSION (active)

Running Water	Glacial Ice	Wind
force	of	flow
(hydraulic action)	(exaration)	(deflation)
corrasion	corrasion	corrasion
attrition	attrition	
cavitation	plucking (quarrying)	
solution	gouging (bulldozing)	
corrosion		

Expansion and contraction are most effective in a rock with four or five minerals, as the change of shape with temperature will vary for each. *Exfoliation* is the falling off of spherical shells of rotten, weathered rock. Granite cliffs do this. Running water includes both stream and wave action.

Force of flow is the power of a current sweeping material along. It is the same for all three types of erosion despite different names being applied. *Corrasion* is mutual wear between a particle in transit and the bedrock. *Attrition* is mutual wear between two particles in transit. *Cavitation* is the surprisingly powerful effect of bursting bubbles in turbulent flow. Freezing ice will *pluck or quarry*, while wind and water will flow around a projection. *Gouging* by continental glaciation produced the thousands of lakes of the northern United States and Canada, including the Great Lakes.

RELATIONSHIP OF STREAM TYPES AND PATTERNS. The *consequent stream type* develops on and as a consequence of an initial slope and has the same direction and tilt. Its *dendritic stream pattern* is tree-like or root-like. The tributaries come in at shallow angles, pointing downstream. The *antecedent stream type* established its course before deformation occurred. Its tributaries adjust to the structure and are of the *subsequent* type. The whole pattern is trellis. The *superimposed stream type* established its course after deformation. It usually starts as a dendritic consequent stream on overlying flat beds. Cutting down into the folded strata, it superimposes its dendritic course on them.

Trellis drainage patterns occur in folded, parallel, mountainous areas like the Appalachians. The major streams cut through the mountains. The tributaries are subsequent, flow on the weak shales in the valleys, and enter the main stream at right angles. *Parallel drainage patterns* occur on even slopes to the sea. *Rectangular drainage* is shifted by faults raising resistant formations in its path.

(Continued on page 108)

EXPLANATIONS

1. *Geo* means earth, *morphos* means form.

2. Aggradation is the raising of the grade, slope, or profile by deposition (not by mountain building or uplift).

3. In weathering, the material is not transported, so the process is passive. In erosion, the material is transported so the process is active.

4. Corrasion is mutual wear between a particle in transit and the bedrock. Attrition is mutual wear between two particles in transit.

5. Solid glacial ice performs gouging or bulldozing; water and air do not.

6. Even in a desert, the most material is moved by running water in times of flash flood.

7. A stream newly formed on an initial surface (without control by the rock structure) develops a dendritic (tree like or root like) shape where the tributaries all come in at shallow angles all pointing downstream.

8. An antecedent stream carved its valley before the deformation occurred.

9. The trellis drainage pattern is controlled by the rock structure. In parallel, folded mountain areas, the tributaries are cutting into the weak shales in the valleys and come into the main stream at right angles. The main streams cut across the mountains and form the "upright" laths of the trellis pattern.

10. Rectangular drainage patterns are caused as faults uplift resistant formations in a stream's path, causing it to shift laterally.

11. Youthful streams carve slots in solid, hard rock such as granite. Such slots are widened in softer sedimentary formations by slump, mass wasting, etc. to a *V*-shaped cross section.

12. Water covers the V-shaped valley bottom or floor in the stage of youth.

13. A stream deviates from a straight path and meanders as it starts to cut its bank laterally when going into the stage of maturity.

14. The stream is no longer spending all of its energy downcutting, but is cutting laterally as well and beginning to deposit sediment.

15. In the mature stage of regional development, the land surface is mostly in slope. There is very little flatland either on the divides or in the valley bottoms.

16. A longshore current travels parallel to, or along, the shore line.

17. A bay barrier bar is a sand bar built across the mouth of a bay, as a barrier by the longshore current.

18. A playa lake is shallow and has very little water most of the year. In the rainy season, however, it expands in size and depth rapidly.

19. The barchan is a crescent-shaped sand dune with horns pointing downwind; in a parabolic, the horns point upward.

20. Loess is composed of silt sized particles that form unstratified, unsorted, thick deposits in the lower valleys of major streams. It is generally thought to be wind deposited. It stands in road cuts as vertical banks without eroding to a slope.

Answers

geomorphology	1
degradation	2
transported	3
attrition	4
glacial ice	5
running water	6
dendritic	7
antecedent	8
trellis	9
rectangular	10
V	11
water	12
meander	13
downcutting	14
slope	15
longshore	16
bay barrier	17
playa	18
barchan	19
loess	20

superimposed) and tributary streams (subsequent). *Stream Patterns* may be either initial slope (dendritic) or structure controlled (trellis, parallel, rectangular, circular, centrifugal and centripetal).

LAND FORMS OF RUNNING WATER. Fluvial (stream) regions are *classified*, like people, *into youth, maturity,* or *old age stages*, based on changes in land forms.

Single valley development involving a single stream is classified by changes in the long profile, the cross section, the map view, and the action performed.

Regional valley development involves many streams in many valleys.

Stream piracy occurs when a stream "beheads" another stream by headward erosion and captures the drainage from the upper part of the other stream's basin.

LAND FORMS OF WIND. *Mountainous rocky areas* are characterized by the following: *Playa lakes* are shallow desert lakes that are dry or greatly reduced most of the year, but expansively wide in the rainy season. *Pediments* are barren rock slopes between the desert playa lake and the mountain crest. The sandy soil has been swept away by the wind. *Ventifacts* are pebbles, sand blasted by the wind.

Sandy areas (shores and deserts) are characterized by the following: *crescent shaped dunes* may be *barchan* (the horns point downwind) or *parabolic* (the horns point upwind). *Long sand ridges* may be *longitudinal* (parallel to prevailing wind direction) or *transverse* (at right angles to the wind). *Frosted sand*, in which the grains are "sandblasted" by wind-driven mutual erosion, giving a surface like a frosted light bulb.

LOWER RIVER VALLEY AREAS are characterized by *loess* (unsorted, unstratified thick deposits of silt-sized particles, probably windblown), which stands in vertical cliffs without slump or slope.

SINGLE VALLEY DEVELOPMENT. The *youthful stream* has a steep grade or profile, hence it is vigorously downcutting and cuts a straight line path across anything in its way. Its cross section is a slot in solid rock (for example, Royal Gorge, Colo.) but widens to a V-shape in softer formations by gravity actions such as mass wasting, slump, and creep. The gentler sloped *mature stream* is doing more lateral cutting and hence tends to widen its valley and spread out. Its path begins to meander. The meander bends touch the valley walls on each side. The cross section shows a flat valley floor (first by erosion, later by deposition), with lower but still definite valley walls. Water does not cover the entire valley floor.

The *old age stream* is flowing over a flat almost featureless plain. Deposition exceeds lateral cutting. The tight, almost closed, meanders may be cut off forming a chute and leaving an abandoned meander bend that may be filled with water to form an oxbow lake. Typical features are natural levees, flood plains, ill-defined valley walls, deltas, and distributaries. The delta has top and bottom set beds of fine material and fore set beds, between, of coarse material.

The greatest force of flow and the deepest erosion of the channel is in the middle on a straight stretch. On a curve, it is on the outside edge where it produces an *undercut bank*. On the inside edge, deposition builds up sandbars. Rain slips down the gentle *slip-off slope* here into the stream. The undercut banks constantly wear away, and so the meander pattern expands sideways, cutting the valley walls and widening the valley.

REGIONAL VALLEY DEVELOPMENT. In *youth*, the divides are broad flat plateaus; a few streams cut deeply downward increasing the relief. In *maturity*, the divides are narrow and rounded; the land surface is mostly in slope and hilly; the number of streams and the amount of relief are both at a maximum. In *old age*, the divides are low, wide apart, and ill defined; the land surface is almost a flat featureless plain of low, decreasing relief; a few streams meander broadly and build depositional features on the wide valley bottoms.

LANDFORMS OF WAVE ACTION. A *bay mouth bar* is formed when a longshore current (one that meets the coast at an angle instead of head on) transports material parallel to the shore, as well as eroding. It builds beaches (mostly sandy) and sweeps sand out across the mouth of a bay or inlet to build a bay mouth bar. A *hook or spit* is produced by tidal currents curving the end of the bar inland. A *back swamp lagoon* is produced by vegetation and stream-fed mud and slit filling up the bay behind the bar. The shore line is thus straightened out. An *offshore barrier bar* is produced where wave base (the depth to which wind-produced waves disturb the water) touches the bottom. Here the breakers are formed (surf zone). Sand piles up and eventually builds above the surface as an offshore barrier bar.

28   GEOMORPHOLOGY—GLACIERS

SELF-TEST

1. Valley glaciers are tributary to _____ _____ glaciers.

2. The top surface of a glacier is called the zone of _____.

3. The half-bowl-shaped depression at the head of a mountain glacier is called the _____.

4. A pyramidal shaped mountain peak left by glaciers eroding back into it from 3 or more sides is called a _____.

5. Valley glaciers do not carve their floors as deeply as the main glacier they feed and hence leave _____ valleys.

6. Glaciers carve _____ shaped valleys (in cross section).

7. The ridge of debris dumped at the forward edge of a stagnating glacier is called a _____ moraine.

8. The merging of the inside lateral moraines of two tributary valley glaciers forms a _____ moraine of the main glacier.

9. Loose, unconsolidated, unsorted, ungraded, non-bedded material dumped by the glacial ice is called _____.

10. Loose, unconsolidated, sorted, graded, bedded material deposited by the meltwater runoff is called _____.

11. A streamlined hill, having the form of an inverted bowl of a teaspoon, depositional in origin, composed of loose, unconsolidated glacial till is called a _____.

12. A depression in the outwash plain left by blocks of stagnating melting ice is called a _____.

13. A drowned glacial valley is called a _____ _____.

14. Northern Hemisphere continental glaciation occurred in the _____ Epoch of the Cenozoic Period.

15. Southern Hemisphere continental glaciation occurred in the _____ Period of the Paleozoic Era.

16. This Southern Hemisphere glaciation is simplified if the theory of Continental _____ is correct.

17. Worldwide temperature change sets the stage for continental glaciation. One thing that might affect this is a change in thick insulating cloud blankets. This is known as the "_____ effect."

18. Since continental glaciation was widespread but not worldwide at any one time, more localized events must have added their effects to the worldwide temperature change to produce this glaciation at spasmodic intervals. More localized causes would be changes in atmospheric and oceanic _____.

19. The many ideas advanced as causes of continental glaciation illustrate the use of _____ Working Hypothesis.

1 _____
2 _____
3 _____
4 _____
5 _____
6 _____
7 _____
8 _____
9 _____
10 _____
11 _____
12 _____
13 _____
14 _____
15 _____
16 _____
17 _____
18 _____
19 _____

Geomorphology—Glaciers

BASIC FACTS

GLACIERS are classified as mountain glaciers or continental glaciers. *Mountain glaciers* are either *piedmont* (large, main glaciers) or *valley* (small tributary glaciers) glaciers. A *continental glacier* is vast in size. It overrides all but the highest peaks.

STRUCTURE OF A GLACIER. In the horizontal areas, there is accumulation and wastage. The vertical zones are the flow (base) and the fracture (top). In the zone of flow at the base of a glacier, no cracks can exist due to the pressure (weight) of the ice above. At the top, *crevasses* are numerous.

LAND FORMS OF GLACIAL ICE. Mountain glaciers and continental glaciers are *erosional forms*. *Mountain Glaciers* are characterized by cirques, cols, horns, arêtes, hanging valleys, tarns, paternoster lakes, U-shaped valleys, striae, rock polish and rock flour. *Continental Glaciers* are characterized by roche-moutonnée, fiords, striae, rock polish, and rock flour.

Depositional Forms may be: *ice-deposited*, moraines (terminal, lateral, medial, and ground) and drumlins composed of unsorted till, or *meltwater deposited*, eskers, kames, kame terraces, rock flour, and outwash plains composed of sorted drift. Kettle holes are numerous.

There have been multiple occurrences of continental glaciation in northern parts of North America and Eurasia, and in central portions of South America and southern portions of Africa, India, and Australia. This glaciation has occurred 1 million to 12,000 years ago (Northern Hemisphere) in the Pleistocene Epoch, 225 to 270 million years ago (Southern Hemisphere) in the Permian Period, and 600 million years ago in Precambrian times. (The theory of

(Continued on page 112)

ADDITIONAL INFORMATION

LAND FORMS OF GLACIAL EROSION. A *cirque* is a half-bowl shaped depression at the head of a mountain glacier. A *col*, or saddle, is formed where two headward eroding cirques meet at the crest of a divide. Where three or more cirques carve back into a mountain, they form a pyramidal *horn*. *Arêtes* are sharp-peaked side ridges between adjacent glaciers. After the ice melts, the valley floor of the smaller valley glacier (which didn't cut as deeply) is left above the floor of the main Piedmont glacier valley. This forms a tributary *hanging valley*, sometimes with spectacular waterfalls. The basinlike steps left by successive cirque positions in the profile of a glaciated valley become filled with water as the ice melts. These glacial-remnant lakes are called *tarns*. A chain of them connected by streams (looking like a string of beads on a map) are called *paternoster lakes*. Glaciers carve the valley walls as well as the floors, and so form *U-shaped valleys*. Striae are deep scratches left in bedrock by glacial passage. They indicate the direction of ice movement. This also causes *rock polish* and the finely ground up *rock flour* that colors the streams white.

LAND FORMS OF GLACIAL DEPOSITION. The *roche-moutonnee* is erosional and composed of solid rock. The *drumlin* is depositional and composed of loose, unsorted till. The former is planed off, the latter is plastered over, a projecting spur. Both have the same streamlined form, like the inverted bowl of a teaspoon.

Fiords are drowned glacial valleys. *Medial moraines* of Piedmont glaciers are formed by the merger of the inside lateral moraines of valley glaciers.

Kames are irregular meltwater runoff deposits. They may form depositional terraces. *Kettle holes* are depressions left in the outwash plain by blocks of melting ice.

CONTINENTAL GLACIATION: PROBABLE CAUSES.

Worldwide lowering of temperature. Some organisms live only in an environment of restricted temperature range. The study of the fossil content of a bed may thus indicate the temperature of the environment in which the formation was deposited. These fossil paleotemperature studies and lowered snowfall lines indicate a worldwide lowering of temperature (2°–10° C). Large amounts of volcanic dust in the atmosphere from explosive activity may cut down the amount of sunlight that gets through to us, thus reducing temperature. Large-scale volcanic activity did precede the Pleistocene Epoch. Also, removal of dense cloud layers would reduce temperature.

Raising of temperature. Conversely, the formation of dense, thick cloud layers would tend to raise temperatures by the greenhouse effect. The greenhouse effect is the trapping

(Continued on page 112)

EXPLANATIONS

1. Piedmont glaciers are fed by tributary valley glaciers.

2. The zone of fracture (crevasses) is at the top surface of a glacier.

3. The cirque depression is at the head of a mountain glacier where it erodes back headward into the mountain.

4. Horns are named after the most famous example in the Swiss Alps, the Matterhorn.

5. Valley glaciers leave *hanging* valleys whose floors are high above the floor of the main Piedmont valley.

6. Glaciers carve the walls as well as the floors and hence leave U-shaped valleys in cross section.

7. A terminal moraine is at the forward edge of a glacier. This marks how far it advanced.

8. Medial moraines of the main Piedmont glacier show as dark lines of material on its surface, merging from separate tributary valley glaciers.

9. Till is scrambled, ice-dumped debris.

10. Drift is bedded, meltwater-deposited debris. The water does a better job of sorting, grading, and bedding.

11. A drumlin is depositional and composed of loose till. A roche-moutonnée is erosional and composed of solid rock. Both have the same streamlined form.

12. A kettle hole is left by melting ice blocks.

13. A fiord is a steep-sided, deep, drowned glacial valley.

14. The Pleistocene Epoch was the time of glaciation in northern North America and Eurasia.

15. The Permian Period saw Southern Hemisphere glaciation.

16. Continental Drift would simplify the widespread areas of Southern Hemisphere glaciation by considering South America, Africa, India, the Antarctic, and Australia as one great land mass.

17. The "greenhouse effect" raises the temperature beneath thick insulating cloud blankets by trapping heat. Ultraviolet rays come in but infrared heat rays can't get out.

18. Changes in atmospheric and oceanic currents were more localized causes of glaciation.

19. Multiple Working Hypothesis is a principle applied to continental glaciation, continental drift, and other things for which no one definite specific answer can be given.

Answers

Piedmont	1
fracture	2
cirque	3
horn	4
hanging	5
U	6
terminal	7
medial	8
till	9
drift	10
drumlin	11
kettle hole	12
fiord	13
Pleistocene	14
Permian	15
drift	16
greenhouse	17
currents	18
multiple	19

Southern Hemisphere glaciation is greatly simplified if all the continents were all together in one great land mass at the time. (See Chapter 31.)

PROBABLE CAUSES OF CONTINENTAL GLACIATION. Since glaciation was widespread, but not worldwide, two different requirements may be necessary:

Worldwide change in temperature. Lowering the temperature would reduce summer melting. Raising the temperature would give increased precipitation. These could result from changes in either the sun (output or distance) or our atmosphere (composition or thickness).

Simultaneous localized changes in atmospheric and oceanic currents. These could be caused by: changing of the tilt of the earth's axis of rotation, shifting of rigid crust over more mobile interior (axis unchanged), or upraising of mountains or undersea ridges.

It should be kept in mind that these must account for several periods of glaciation.

of heat and raising of temperature by thick insulating cloud layers. Shorter wavelength ultraviolet rays penetrate these layers, but the resulting longer wavelength infrared heat rays can't get back out. If the volcanic activity were more eruptive than explosive, large quantities of water vapor would be given off. Some geologists say a rise in temperature giving increased evaporation from the sea is the only way to achieve the quantity of precipitation needed. To the south of the Pleistocene ice sheets, excessive rainfall created huge pluvial lakes in the southwestern United States. The Great Salt Lake is but a small remanant of the former Lake Bonneville. The Sahara Desert area was green, fertile, and well populated at this time.

Tilting. The geological history of the earth indicates uniform widespread climates over most of the earth during the last 600 million years, except for the large-scale glacial periods mentioned. Thus it is possible that throughout much of geological time the earth's axis may not have been tilted and thus there were no seasons.

Paleomagnetism (plotting the direction and dip of ferromagnesian crystals "frozen" in the solidified magma) indicates considerable wandering of the magnetic poles. The geographic poles may or may not have moved with them. If they did not, but the rigid outer part of the earth shifted over a more mobile inner layer, different points would be brought under the poles at different times.

Uplift of Mountains or Ridges. The cutting off of the interior of North America from moisture-laden sea winds by the upraising of the Appalachian and Rocky Mountains lowered the temperatures considerably. Continental elevations rose generally in the Cenozoic Era.

Europe has the same latitude as northern Canada, but the warm Gulf Stream gives it the climate of the United States. A change in course of this ocean current would profoundly change the climate of the Northern Hemisphere.

Comparable Glacial Land Forms

	Ridges	*Blankets*	*Streamlined Hills*
Erosion	arête	rock polish	roche-moutonnee
Ice deposit	moraine terminal lateral medial	moraine ground	drumlin
Water deposit	esker	outwash plain	

29 STRUCTURAL GEOLOGY

SELF-TEST

1. The movement of molten rock is called _____.

2. The movement of solid rock is called _____.

3. Molten rock material below ground is called _magma_.

4. Molten rock material at the surface of the ground is called _lava_.

5. The massive, subjacent, intrusive, igneous, rock form that wells up from underneath is called a(an) _batholith_.

6. The discordant, crosscutting, wall-like, igneous rock form is called a(an) _____.

7. The concordant, sheetlike, igneous, intrusive, rock form, intruding between two other layers, is called a(an) _____.

8. The concordant, igneous, intrusive rock form that domes the rocks above and has a flat floor is called a(an) _laccolith_.

9. The extrusive igneous rock form that builds a cone is called a(an) _____.

10. The extrusive igneous rock form that spreads out over extremely large areas, but does not build a cone, is called a(an) _____.

11. Lava that cools in ropy, strandlike coils is called _____.

12. An upwarping of a series of strata along a long axis is called a(an) _____.

13. A downwarping of a series of beds to a central point is called a(an) _____.

14. A downwarping of a series of beds along a long axis large enough to be plotted on a globe or continental map is called a(an) _geosyncline_.

15. A separation (in the rocks of the earth) involving frictional movement of one wall over another is called a(an) _fault_.

16. "The hanging wall appears to be downthrown," describes a _normal_ fault.

17. A low-angle reverse fault, usually with great displacement, is a sign of a(an) _thrust_ fault.

18. In the primary wave, the direction of motion of the particle is _____ to the direction of motion of the wave.

19. The point on the surface of the ground directly above the center of earthquake activity is called the _epicenter_.

20. An instrument for recording earthquake waves is called a(an) _seismograph_.

1. _____
2. _____
3. _____
4. _____
5. _____
6. _____
7. _____
8. _____
9. _____
10. _____
11. _____
12. _____
13. _____
14. _____
15. _____
16. _____
17. _____
18. _____
19. _____
20. _____

BASIC FACTS

STRUCTURAL GEOLOGY deals with the shape and form that rocks assume under stress from the active deforming agents within the earth's crust. Two major processes take place—vulcanism and diastrophism. *Vulcanism* is the movement of molten rock. *Diastrophism* is the movement of solid rock.

Vulcanism consists of intrusion and extrusion of magma or lava. *Magma* is melted rock material below the ground. *Lava* is melted rock material at the surface or the solidified rock formed from it.

In *intrusion*, displaced magmas are raised from lower levels to higher levels. In *extrusion*, displaced magmas are raised from lower levels to the surface. In either case *displacement* is taking place, usually upward (easiest direction of relief of pressure).

Melting is thought to take place at depths of 20 to 50 miles, when 2 or more magma melting factors contribute to raise the temperature above the melting point locally.

MAGMA MELTING FACTORS include: natural geothermal gradient (increase in temperature with depth), heat of friction, the fact that pressure release changes state, and concentration of radioactivity.

INTRUSIVE ROCK FORMS include: *subjacent*, which wells up beneath strata (batholith); *discordant*, which cuts across strata (dike, neck); and *concordant*, which intrudes between strata (sill, laccolith).

EXTRUSIVE ROCK FORMS include: *volcanoes* (eruptions from a vent building a cone; the types vary in composition of magma, explosiveness, material ejected, and steepness of slope) and *fissure flows* (vast outpourings of basaltic lava from a crack or fissure without building a cone).

DIASTROPHISM is the movement of solid rock. Rock that is deeply buried behaves plastically and will fold or bend under pressure.

(Continued on page 116)

ADDITIONAL INFORMATION

INTRUSIVE ROCK FORMS. A *subjacent type* rock form wells up underneath. A *batholith* melts its way upward; it is massive and enlarges downward. A batholith has no visible or inferable floor and does not dome the rocks above it. It feeds the other forms. Batholiths are composed of granite rock.

A *discordant type* rock form cuts across strata. A *dike* is a wall-like intrusion that stems from a batholith. A dike is not as wide as it is long and high, and is usually basalt. A *neck* is a circular dike that reaches the surface and feeds a volcano.

A *concordant type* rock form intrudes between strata. A *sill* is a sheet-like intrusion sandwiching between beds of rock along a zone of weakness. Sills are fed by magma from a dike and are usually basalt. A *laccolith* is a half-lens-shaped intrusion with a flat floor, formed where magma from a dike meets resistance in penetrating as a sill between a bedding plane, piles up, and domes the rock above it. A laccolith is usually basalt.

EXTRUSIVE ROCK FORMS. *Volcanoes* are formed by the eruption of lava at the surface, building a cone. They may be listed as active, passive, or extinct according to the recentness of eruption. The classic volcanic types (in order of increasing violence of eruption) are: Hawaii, Stromboli, Vulcan, Pelee, and Krakatoa.

Range of Volcanic Types

	Hawaii	Krakatoa
Composition of magma	basic	acid
Explosiveness	passive	extreme violence
Material ejected	free flowing lava	fragments and ash
Steepness of slope	low	steep
	(shield cones)	(cinder cones)

Two main areas of volcanic activity exist:

The circum-Pacific chain (the "Ring of Fire") includes the Andes Mountains, the Coast and Cascade Ranges, the mountains of Western Canada and Alaska, the Aleutian Islands, Japan, and the South Sea Islands; it encircles the Pacific with volcanoes. The Mediterranean Sea and the Caribbean Sea areas have numerous volcanoes.

Types of Lava: Aa—very rough surface; tends to cool in square blocks. Pahoehoe—surface appearance is like coiled strands of rope, highly viscous and stringy. Columnar—cools to form hexagonal prisms or pillars. Pillow—elliptical forms, usually produced by cooling beneath the surface of marine waters. Tuff—volcanic ash and cinders, blown into the air, settled, and sorted by wind and running water into bedded deposits.

TYPES OF FOLDED STRUCTURES: *Homocline*—a series of strata all tilted at the same homogeneous inclination. *Monocline*—one inclination, or one steplike bend, in an otherwise

(Continued on page 116)

Structural Geology

EXPLANATIONS

1. Vulcanism is one of two major processes of change.
2. Diastrophism is the other major process of change.
3. The intrusion and extrusion of magma and lava constitute vulcanism.
4. Lava is molten rock at the surface, or rock that has cooled at the surface.
5. A batholith is massive. It melts its way upward, wells up from underneath, enlarges downward, and has no visible or inferable floor.
6. A dike is not very wide, but long and high. It stems from a batholith and feeds all the other igneous rock forms. It cuts across sedimentary rock beds and is hence discordant.
7. A sill is not very thick, but long and wide. It intrudes between sedimentary rock beds and is hence concordant.
8. The laccolith is fed from the batholith by a dike feeder. It intrudes between sedimentary rock beds and hence is concordant.
9. A volcanoe may be active, passive, or extinct.
10. A fissure flow spreads extrusive igneous rock out over very large areas (for example, 100,000 sq mi in India) without building cones.
11. Lava may assume various forms.
12. An anticline is an upwarping of a series of strata along a long axis. The beds dip away from the axis on each side.
13. A basin is a downwarping of a series of beds to a central point. The beds dip downward into the central point from all sides.
14. A geosyncline is a syncline large enough to be plotted and visible on a globe or continental size map.
15. A fault is a separation involving frictional movement of one wall over another. A joint does not involve frictional movement; it is a mere pulling apart.
16. In a normal fault, the hanging wall (the wall hanging over the fault plane) appears to be downthrown.
17. The thrust fault is under 45° (usually 15° to 25°) of slope. The hanging wall is thrust up over the foot wall, and thus it is a reverse type fault.
18. The primary wave is longitudinal and therefore the particle motion is parallel to the wave motion.
19. The epicenter is on the surface and lies directly above the focus which is the subsurface center of fault movement producing the earthquake shocks.
20. The seismograph is an instrument that records seismic waves, the shock waves resulting from movement of rock along a fault.

Answers

vulcanism	1
diastrophism	2
magma	3
lava	4
batholith	5
dike	6
sill	7
laccolith	8
volcano	9
fissure flow	10
pahoehoe	11
anticline	12
basin	13
geosyncline	14
fault	15
normal	16
thrust	17
parallel	18
epicenter	19
seismograph	20

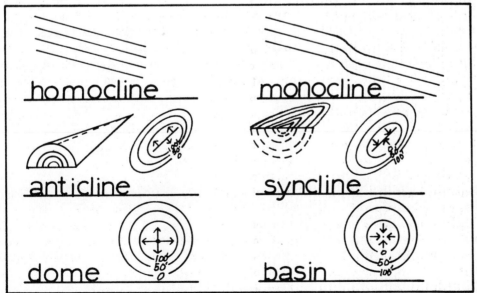

FIG. 29.1. Types of Folded Structures.

FOLDED STRUCTURES include: homoclines, monoclines, anticlines, domes, synclines, geanticlines, and geosynclines (see Figure 29.1). Rock will fold until it reaches its breaking point. Here it will break or separate.

SEPARATION STRUCTURES include: *joints* (separations, in the rocks of the earth, with no frictional movement of one wall over another; a mere pulling apart) and *faults* (separations, in the rocks of the earth, involving frictional movement of one wall over another).

ATTITUDES OF STRUCTURES are shown by: the *strike* (the trend or compass direction of the intersection of an inclined structure with the horizontal) and the *dip* (the angle between the plane of an inclined structure and the horizontal). Dip is, always measured in a vertical plane at right angles to the strike.

TYPES OF FAULTS. A geometric classification is commonly used based on apparent movement of the hanging wall. This gives normal, reverse, thrust, and vertical faults.

EARTHQUAKES are shock waves felt at the surface of the ground caused by displacement of rock along a fault.

Types of earthquake waves include:
Primary waves, which travel fastest and arrive first. Primary waves travel through the earth and move in both solids and liquids. They are longitudinal waves. *Secondary waves* travel slower and arrive second. These waves also travel through the earth but can move only in solids. They are transverse waves. *Long waves* travel slowest and arrive last. These waves travel the longest path along the surface of the earth. They are complex waves.

The difference between the arrival times of primary and secondary waves tells the distance away from the origin of the earthquake.

flat series of strata (a local steepening of dip). *Anticline*—an upwarping of a series of strata along a long axis. *Dome*—an upwarping of a series of strata to a central point. *Syncline*—a downwarping of a series of strata along a long axis. *Basin*—a downwarping of a series of strata to a central point. *Geosyncline (geanticline)*—a syncline or anticline large enough to be shown on a continental map or globe (may be 1000 to 2000 miles long). *Plunge*—the angle at which the inclined axis of a tilted anticline or syncline dips into the ground.

FAULT TERMS: *Hanging Wall*—the wall that hangs over or is above the fault surface. *Foot Wall*—the wall that lies beneath the fault surface. (These terms were developed in Germany in the early days of mining there.)

TYPES OF FAULTS (geometric classification): *Normal fault*—the hanging wall appears to be downthrown. *Reverse fault*—the hanging wall appears to be upthrown. *Thrust fault*—a low-angle reverse fault, usually with great displacement. *Vertical fault*—the fault surface is straight up and down and movement is usually sideways along the strike of the fault (strike-slip). The terms hanging and foot walls do not apply here.

A normal fault exhibits gap. A reverse fault exhibits overlap. These are usually not more than one-half mile wide. The overlap of a thrust fault, however, may be 100 miles or more wide, and the edges of the beds are drag-folded backwards. The angle of dip of normal and reverse faults is high (between 45° and 90°). The thrust fault is under 45° (usually 15° to 20°).

The reason for preferring a geometric classification is its simplicity. Only one point on the same bed on each side of the fault has to be known. To use a genetic classification (tension faults, compression faults, gravity faults, etc.), one should prove which side moved up and which went down (or both).

30 MOUNTAINS AND CONTINENTS

SELF-TEST

1. The folded geosyncline type of mountain is typified by the
   a. Appalachians ✓
   b. Alps
   c. Rockies

2. The folded and cross-faulted type of mountain is exemplified by the
   a. Appalachians
   b. Alps
   c. Rockies ✓

3. An example of the complexly-thrust faulted type would be the
   a. Appalachians
   b. Alps ✓
   c. Rockies

4. An example of horst and graben structure would be the
   a. Sierra Nevada
   b. Basin and Range Province Mts. ✓
      (in Utah and Nevada)

5. An example of the tilted hinged-block fault type of mountain would be the
   a. Sierra Nevada ✓
   b. Basin and Range Province Mts.
      (in Utah and Nevada)

6. An example of the volcanic type of mountain chain would be the
   a. Ozarks
   b. Grand Canyon Region
   c. Andes ✓

7. An example of the central domal uplift type would be the
   a. Ozarks ✓
   b. Grand Canyon Region
   c. Andes

8. An example of plateau uplift would be
   a. Ozarks
   b. Grand Canyon Region ✓
   c. Andes

9. The backbone of the continent is the continental
   a. shelf
   b. platform
   c. shield ✓
   d. slope

10. The part of the platform under the ocean is the continental
    a. shelf ✓
    b. platform
    c. shield
    d. slope

11. The real outer margin of the continent is the continental
    a. shelf
    b. platform
    c. shield
    d. slope

12. The ever-deepening wedge of sedimentary rock is the continental
    a. shelf
    b. platform
    c. shield
    d. slope

13. The continents are formed of an outer layer of light weight rock of the crust that is called the
    a. sima
    b. sial

14. The continents grow by the addition of
    a. marginal geosynclines and seaward borderlands.
    b. seamounts raised from the ocean floor.
    c. slow continued decrease in sea level.

1 ___
2 ___
3 ___
4 ___
5 ___
6 ___
7 ___
8 ___
9 ___
10 ___
11 ___
12 ___
13 ___
14 ___

Mountains and Continents

BASIC FACTS

Types of Mountains.
1. *Folded Geosynclines* (Appalachians and Ouachitas).
2. *Complexly Thrust-Faulted Geosynclines* (Alps and Himalayas).
3. *Folded and Cross-Faulted Geosynclines* (Rockies).
4. *Tilted Hinged Block Faults* (Sierra Nevada).
5. *Horst and Graben Structures* (Basin and Range Province Mts.).
6. *Volcanic Chains* (Andes).
7. *Central Dome Uplifts* (Ozarks and Adirondacks).
8. *Plateau Uplifts* (Grand Canyon Region).

Structure of the Continents.
A *continental shield* is large centrally-located segment composed of old igneous and metamorphic rocks. A *continental platform* is an ever-deepening wedge of sedimentary rock running from the shield into the ocean. The *continental shelf* is the outer part of the platform that is under the ocean. The *continental slope* is the steep escarpment at the outer edge of the shelf. The *continental rise* is a leveling off to the flat sea floor. The *abyssal plain* is the flat bottom of the ocean floor.

Composition of the Earth's Crust.
The outer layer of silicon-aluminum. Lighter-weight rock making up the mass of the continents is called *sial*. *Sima* is the base of heavier-weight silicon-magnesium rock underlying the continents and flooring the ocean basins.

The Igneous Cycle.
1. Formation of a geosyncline, to a thickness of 50,000 ft.
2. Formation of a magma pool at depths of 10 to 20 miles.

(Continued on page 120)

ADDITIONAL INFORMATION

The Major Types of Mountains. The geosyncline is a long, relatively narrow, marine seaway along the margin of a continent. It receives sediment eroded off the continental interior and also from a borderland seaward of it. The base of the sediment in the geosyncline sinks into the crust below at a rate equal to the deposition. A shallow (never deep) seaway is always maintained at the surface overlying sediment thicknesses of 50,000 ft. Later, lateral compression of the borderland toward the continental mass produces *folded geosyncline mountains* like the Appalachians. Such a surface may rise (from 200 ft below sea level) to 14,000 ft above sea level. Long (800 to 1000 mi) anticlines alternate with synclines. Physiographically, these regions are called ridge and valley provinces.

Especially in the *Alps*, the most striking thing is the massive thrust faulting that occured after the initial folding. Marine formations once part of the Mediterranean Sea botton have been thrust up as mountains 75 to 100 miles over younger beds.

The *Rocky Mountains* are essentially a long geosyncline like the Appalachians, but they differ by being "chopped up" into many short ranges and intermontane basins by east-west crossfaults. They are thus much more complex.

The *Sierra Nevada* Mountains are a massive 1000 mile long block, hinged at the west side, and tilted with a great fault scarp at the eastern edge. The western slope is gentle while the eastern escarpment slope is quite steep.

Horsts are high blocks on either side of a down-dropped block that is called a *graben*. The Utah-Nevada area, as a geographic, physiographic province is called the *basin and range province*. Here the flat-topped mesa mountains are horsts and the valleys are down-dropped grabens.

The *Ozarks* have folding and faulting, but the most prominent thing is the doming at the center.

In the *Grand Canyon region*, beds have been uplifted nearly a mile with little deformation. They are still flat-lying and nearly horizontal.

Continental shields are old igneous and metamorphic rocks that form the backbone of every continent (Asia has two). They have been land areas (not marine) through most of geologic time, and have little, if any, cover of sedimentary

(Continued on page 120)

EXPLANATIONS

1. Folded geosynclinal mountains are formed by lateral compression of the seaward borderland toward the landward continental mass as a result of pressures caused by sediment deposition and/or convection cell movement.

2. The Rockies are essentially a long geosyncline like the Appalachians, but they are broken by east-west crossfaults.

3. The Alps are the result of massive thrusts after initial folding.

4. Horsts are high blocks on either side of a down-dropped block called a graben.

5. The Sierra Nevada are hinged at the west side and tilted with a great fault scarp at their eastern edge.

6. The Andes are mostly the result of volcanic action.

7. The Ozarks also have folding and faulting, but doming is the most prominent feature.

8. In the Grand Canyon, beds have been uplifted nearly a mile, with almost no deformation.

9. Continental shields are composed of old igneous and metamorphic rocks.

10. The continental shelf is the outer part of the continental platform.

11. The continental slope is the steep escarpment at the outer edge of the continental shelf.

12. By definition.

13. The heavier rock underlying the continents and flooring the ocean basins is called sima.

14. Accretion of these marginal seaward features adds to the bulk of the continents.

Answers

a	1
c	2
b	3
b	4
a	5
c	6
a	7
b	8
c	9
a	10
d	11
b	12
b	13
a	14

3. Extrusion of vast fissure flows from the sima.
4. Extrusion by volcanic cones from the sial.
5. Diastrophic folding and upraising of the geosyncline into folded mountains.
6. Emplacement of the granite batholith at the roots of folded mountains.
7. Intrusions of sills, laccoliths and veins from the batholith into the country rock.
8. Dying remnants of vulcanism (hot springs and geysers) appear as water table is re-established.

GROWTH OF CONTINENTS. Continents grow as marginal geosynclines, and their seaward borderlands are added to the continental mass. The borderlands may be chains of island arcs as Japan and the Kuriles are today, feeding sediment westward into the geosyncline of the Japan Sea. Continents may also grow as the marginal slopes and rises are accordion-pleated inward by ocean floor convection, without seaward borderlands actually existing.

rocks deposited over them. The *continental shelf* is the submerged part of the platform between the shore line and 100 fathoms (600 ft) of water depth. The *continental slope* is the outer edge of the continental platform (shelf). The slope steepens abruptly and descends thousands of feet within a few miles to around 1700 fathoms (10,200 ft). This is the real outer edge of the continent. The *continental rise* is a lesser grade between the continental slope and the flat *Abyssal plain* at 3000 fathoms (18,000 ft).

THE IGNEOUS CYCLE AND THE GROWTH OF CONTINENTS. It is by means of the igneous cycle that continents grow in extent. The borderlands that lie seaward of the geosynclines are considered today to be *volcanic island arcs*. An *outer eugeosynclinal zone* has volcanic material while an *inner miogeosyncline* does not. The combined shield and platform mass is called a *craton*.

The increasing weight of sediment in the geosyncline causes melting of the crust below it. Here, at the margin of the continent, the sima layer (and the mantle) lie at lesser depths. If the magma pool forms in the sima, plateau basalts are extruded. If it forms in the overlying sial, volcanic andesites to rhyolites are extruded.

Isostasy refers to a balance between upward-dwelling forces and downward-dwelling forces. As the geosyncline sinks, convection cell movement transfers material laterally from under it to beneath the borderland causing it to rise. Erosion and deposition complete the cycle.

The extrusions from the magma pool break this isostatic equilibrium. Large volumes of magma are moved upward and, generally, seaward. This mass and the force of its extrusion presses the borderland against the geosyncline (perhaps aided by continental drift). The sedimentary rocks are folded upward into mountain ranges. The release of pressure at the base expands the magma pool through the sial. The surrounding granite melts and flows into the base area. The granite batholiths are emplaced. (Or, since the sedimentary rocks were formed from granite and have the same elements, the melting of them may produce granite.)

A great geosyncline is forming today at the southern margin of the North American continent. New Orleans lies on the central part of the axis of the *Gulf Coast Geosyncline*, which has perhaps 40,000 ft of sedimentary rock beneath.

31

CONTINENTS

SELF-TEST

1. Evidence for continental drift includes: the matching of the _____ of continents, the matching across the same zone of _____, and the matching of _____.

2. The proposed original supercontinent of the Southern Hemisphere has been termed _____.

3. The close of the _____ Era has been cited as a likely time when continental drift may have occurred.

4. Modern opinion today considers the most likely force causing continental drift to be _____ movement.

5. The most probable heat source for this movement is the _____ of minerals at great depth.

6. The most extraordinary feature of the globe-girdling oceanic ridges is that the center lines are downdropped _____.

7. An unexpected finding has been that the sedimentary rock cover of the ocean floor is very _____.

8. Another unexpected finding has been that the ages of the rocks of the ocean floor are very _____.

9. It has been proposed that the Mid-Atlantic Ridge represents the _____ edge of a convection cell, with the graben being a downdropped plug.

10. The outer acidic solid crust of the earth is _____ under the oceans.

11. The generally accepted composition of the mantle is ultra _____ solid rock.

12. Most of our information of the interior of the earth comes from the study of _____ waves.

13. The physical state of the outer core appears to be that of a _____.

14. The fact that _____ waves do not pass through it indicates this.

15. The composition of the core is considered to be approximately 10% nickel and 90% _____.

1 _____
1 _____
2 _____
3 _____
4 _____
5 _____
6 _____
7 _____
8 _____
9 _____
10 _____
11 _____
12 _____
13 _____
14 _____
15 _____

BASIC FACTS

THEORY OF CONTINENTAL DRIFT. The *coastal outlines of the continents* bordering the North and South Atlantic would *fit fairly well* if they were shoved together like a giant jigsaw puzzle. Of course, what should be matched are the outlines at the continental slopes rather than the actual present water contact shorelines. Also, *rock types and fossil zones match* fairly well across this boundary.

These similarities were pointed out in 1915 by Wegener. He went even further to propose that *all the continents may have been together* in one great land mass and have since moved and drifted apart. The southern part of this hypothetical land mass has been named Gondwana land and the northern part called Lavrasia.

The Antarctic continent may have been east of Africa and south of India, with Australia on its eastern coast. Africa thus would originally have had little coast line except for the Mediterranean, a deep sea that apparently has existed for a long time.

The Himalayas and the high Tibet plateau were formed (along with the Rockies and Andes during the Laramide orogeny) by the subcontinent of India pressing upon and overriding the mainland of Asia. The rocks of India match those of Africa and the Antarctic more than those of Asia.

The close of the Mesozoic Era is often cited as *a likely time for continental drift* to have occurred. Such things as widespread uniform climates and widespread similarity of organisms existed prior to this time, but variation in climates and diversity of organisms is found after this point.

(Continued on page 124)

ADDITIONAL INFORMATION

Geologists today generally agree that some amount of continental drift has taken place. The continents are considered to be floating islands of sial moving upon a vast sea of sima.

The International Geophysical Year (1957–1958) investigations showed the Mid-Atlantic Ridge to be but one of many globe-encircling *oceanic ridges*. The top of each of these ridges contains a down-dropped graben—not more than 10 miles wide, but thousands of miles long. The ridges are thought to be upward-welling edges of convection cells. *Convection cell* material heated in the mantle at great depth (perhaps by radioactive decay) expands and its density becomes less. It therefore rises to the surface, spreading out and cooling. As it cools, its density becomes greater and it sinks again. Convection cells thus have upwelling edges on one side and downwelling edges on the other. At the surface, material is moved from the upwelling edges to the downwelling ones, where it is dragged down into the depths.

Current data indicate that: the oldest continental rocks known are $3\frac{1}{4}$ billion years old, the oldest marine rocks known are 60 million years old, the ocean floor cover is surprisingly thin, and that heat loss seems to be approximately equal from continents and ocean basins. The Mid-Atlantic Ridge then may be the upwelling edge of a convection cell with the downwelling edge at the western coast of the American continents, where there are numerous deep-extending faults and fractures. The mountainous ranges of the western parts of these continents represent the crumpling of the continental edges against the convection cell edge, over enormous lengths of time. Two lines of research are currently investigating these problems: *If convection cell movement takes place, the ocean basins should be younger and have less sedimentary* rock cover. This would occur since the ocean floor would be continuously renewed—moved and then dragged under the floating continents. *If radioactive mineral decay at depth is a major heat source* for the convection cell movement, *heat loss should be greater over the ocean basins*. The continents may be moved in order to dissipate heat directly to the ocean.

INTERIOR STRUCTURE OF THE EARTH. The *crust* of the earth may be 10 to 30 miles thick under the continents, but only 3 to 10 miles thick under the ocean floor. Acid granite

(Continued on page 124)

EXPLANATIONS

1. The matching of coastlines was pointed out by Wegner (a German meteorologist) in 1915. *Rock types* and rock structures match across the Atlantic Ocean, as do fossil zones also match.

2. Gondwana land was proposed for the Southern Hemisphere supercontinent and Lavrasia for the Northern Hemisphere one.

3. The close of the Mesozoic Era was proposed (by Wegner) as a likely time for separation of the continents. Some modern geologists suggest the Triassic.

4. Convection cell movement is today most commonly considered to be the force for drift.

5. The probable heat source for such movement was radioactive mineral decay at depth.

6. Long globe-girdling yet narrow (less than 10 miles wide) grabens occupy the tops of the oceanic ridges.

7. Surprisingly, the ocean floor cover has been found to be very thin. Continental drift considers the sediments washed off the continents as drawn down into the interior in constant cycles at the convection cells.

8. In contrast, the rocks of the continents are very old. Radioactive dating records 2 billion years in Canada and Sweden and $3\frac{1}{4}$ billion years in South Africa.

9. The western coasts of the Americas would represent the downdwelling edge of this cell.

10. The crust is thinner under the ocean floor (3 to 10 miles thick) than under the continents (10 to 30 miles thick).

11. The mantle is believed to be mainly composed of olivine with pyroxenes and amphiboles.

12. Earthquake waves furnish most of our knowledge of the earth's interior.

13. The outer core appears to be a fluid.

14. Transverse (secondary) waves do not pass through the outer core. In the laboratory, these waves do not move through a fluid while other types do. Therefore, it is believed that the outer core is a fluid.

15. Three lines of evidence point to this:
    (1) The specific gravity of the earth indicates a very dense core.
    (2) Seismic wave speeds.
    (3) Analogy with meteorite samples (of probable asteroid origin) suggests it.

Answers

coastlines	1
rock types	1
fossil zones	
Gondwana	2
Mesozoic	3
convection cell	4
radio-activity	5
grabens	6
thin	7
young	8
upwelling	9
thin or absent	10
basic	11
earthquake	12
fluid	13
transverse	14
iron	15

These changes argue for movement of the continents and, perhaps, tilt of the earth's axis at this time. (Some modern geologists consider the Triassic a more likely time.)

The cutting off of Australia from the mainland certainly occurred early in the Cenozoic as evidenced by the establishment of primitive mammal forms there, but lack of more recent forms. The disappearance of the dinosaurs and the higher elevation of the continents in the Cenozoic, leading to continental glaciation, are probably tied in with this.

Heat production from radioactive mineral decay causing convection cell movement in the mantle and crust is the most widely supported force proposed for continental drift.

INTERIOR STRUCTURE OF THE EARTH. The earth's outer crust is 3 to 30 miles thick and grades from acid to basic rock going inward. The mantle is approximately 1800 miles thick and is thought to be composed of basic solid rock (peridotite). The outer core is approximately 1380 miles thick and is composed of fluid—melted iron and nickel. The inner core has a radius of approximately 780 miles and is thought to be solid iron (90%) and nickel (10%).

sial grades downward to basic basalt sima. The specific gravity of the earth's crust averages 2.7.

The *mantle* is a massive solid or semi-solid rock layer below the crust. The rocks are not fluid only because of the high pressures there. The specific gravity of the mantle is about 3.3. The rock is probably mostly peridotite, which is olivine with pyroxenes, amphiboles, and micas added.

The *outer core* appears to be a fluid since secondary (transverse) waves are absorbed here and do not penetrate this layer. The average specific gravity of both the outer core and the inner core averages a little over 12.5. The composition of both is assumed to be mostly iron and nickel. In the outer core, this mixture is a fluid at the temperature present there, since the pressure is lower than in the inner core. (The natural geothermal gradient increase is 1° F/60 ft at the surface, but the rate of increase slows going inward.) Convection cell currents in this fluid iron mixture are thought to set up the earth's magnetic field. Changes in its flow account for wanderings of the earths magnetic poles, which may thus move about independently of the geographic pole (axis of rotation) positions.

The *inner core* is assumed to be almost pure solid iron and nickel. There are two reasons for this assumption. Astronomical calculations give the average specific gravity of the whole earth as 5.5; that of the crust is about 2.7. Therefore, the core zones must be much heavier. The specific gravity of iron is 7.86, that of nickel is 8.9. The pressure at the core (about 3 million atmospheres) would actually increase this. The second reason for assuming a 90% iron–10% nickel core is that this is the composition of metallic meteorites, which may be fragments of a planet that once occupied the assorted belt. The stony meteorites have the composition of the mantle.

Seismic waves, the shock waves from earthquakes, give us most of our information about the interior of the earth. Their speed abruptly changes at the boundaries of the zones; hence speeds are discontinuous here and boundaries are called discontinuities. They are named after the men who discovered them. The Mohorovicic discontinuity is the boundary between the crust and the mantle, and the Weichert-Gutenberg discontinuity is the boundary between the mantle and the outer core.

32 FOSSILS AND THEIR CLASSIFICATION

SELF-TEST

1. The remains of a former living animal or plant is called a(an) _____.

2. The entire body of a woolly mammoth is found frozen in the tundra of Siberia. This is a rare case of _____ preservation.

3. Mineral matter in solution directly replaces body tissues. This is an example of preservation by _____.

4. Mineral matter in solution fills in pores and cell spaces in an organism. This is an example of preservation by _____.

5. A carbon film outlining a fossil is left (by high temperatures) on a bedding plane. This is an example of preservation by _____.

6. Molds and casts, gastroliths (gizzard stones), and footprints and trails are _____ traces of fossils.

7. The type of rock most fossils are found in is _____ rock.

8. The environment fossils are preserved best is a(an) _____ environment.

9. A fossil that is restricted vertically in the stratigraphic column and in time but is widespread geographically may be used to identify a formation. It would be called a(an) _____ fossil.

10. External coverings or shells for protection of animals are termed _____.

11. Organisms that look alike and inhabit the same environment, but do not interbreed, are termed _____.

12. Single celled; the one cell performs all body functions; free moving; aquatic. The phylum is _____.

13. Many celled; cells specialize; body sack-like; marine; attached; sponges. The phylum is _____.

14. Animals with true vertebrate columns or backbones; exhibit bilateral symmetry; numerous; wide environmental range. The phylum is _____.

15. A class of the vertebrate animals; live almost entirely in water; extract oxygen from water by means of gills; cold blooded; possess scales; egg laying. The class is _____.

1 _____
2 _____
3 _____
4 _____
5 _____
6 _____
7 _____
8 _____
9 _____
10 _____
11 _____
12 _____
13 _____
14 _____
15 _____

BASIC FACTS

A FOSSIL is the remains of a former living organism (plant or animal). Most fossils are found in sedimentary rocks; usually in marine beds.

Preservation of original body material is quite rare. Most often fossils are the result of *petrifaction* (turning to rock). This occurs either by *replacement* (where mineral matter in solution directly replaces body tissues as they are dissolved by filtering sub-surface waters) or by *permineralization* (the filling in of open pores and cell spaces within the original material by dissolved minerals that precipitate out).

The organism may also leave *indirect traces* such as: molds and casts (although a fossil becomes dissolved, a mold of its shape may be formed; material filling it in later may harden to form a cast), coprolites (fossil body wastes), and footprints or trails.

An *index fossil* is a fossil that identifies a formation. It is restricted in the time of its occurrence (and thus vertically in the rocks), but is widespread geographically. A *fossil assemblage* is a group of fossils that, by their combined overlaping presence, identify a formation.

CLASSIFICATION OF ORGANISMS. Subdivisions (from largest to smallest): kingdom, phylum, class, order family, genus, and species. If necessary, subgroupings are indicated (for example, subphylum).

Genera are organisms that look alike and inhabit the same environment, but do not interbreed. *Species* are organisms that look alike, inhabit the same environment, and interbreed. The generic name comes first and is always capitalized. The name of the species follows it and is never capitalized.

(Continued on page 128)

ADDITIONAL INFORMATION

KINGDOM PLANTAE (food producers—most have chlorophyl—most are immobile) includes:
Subkingdom Thallophyta—one-celled; lack roots, stems, or leaves.
    Algae—one-celled plants with chlorophyl.
    Fungi—one-celled plants without chlorophyl.
Subkingdom Embryophyta—mostly land plants.
    Phylum Bryophyta—small, no roots, includes mosses.
    Phylum Tracheophyta—vascular system distributes sap.
        Subphylum Lycopsida—club mosses and scale trees; coal-forming.
        Subphylum Sphenopsida—rushes; coal-forming.
        Subphylum Pteropsida
            Class Filicinae—ferns; giant tree ferns; coal-forming.
            Class Gymnospermae—ginkgos, cycads, conifers; bare seeds.
            Class Angospermae—hardwood trees and flowering plants; covered seeds.

KINGDOM ANIMALIA (food consumers—no chlorophyl—most mobile) include:
Phylum Protozoa—one-celled.
Phylum Porifera—multicelled; body perforated by pores.
Phylum Coelenterata—multicelled, radial symmetry, tentacles around mouth.
Phylum Platyhelminthes—multicelled, flatworms, bilateral symmetry.
Phylum Aschelminthes—multicelled, roundworms, bilateral symmetry.
Phylum Annelida—multicelled, segmented worms, bilateral symmetry.
Phylum Mollusca—multicelled, tube foot, soft body covered by mantle that usually secretes limy shell, bilateral symmetry.
Phylum Arthropoda—multicelled; external chiton skeleton (exoskeleton), jointed appendages, bilateral symmetry.
Phylum Echinodermata—multicelled, spiney, radial symmetry.
Phylum Chordata—multicelled, internal calcite skeleton, dorsal nerve chord, bilateral symmetry.

Included in the chordates are:

Class Ostracoderm—primitive armored fishes, once considered extinct.

Class Pisces—true fishes, live only in water, cold blooded, lay eggs, have scales, Silurian to present.

(Continued on page 128)

EXPLANATIONS

1. Fossils are the remains of former living organisms.

2. Direct preservation is extremely rare. It sometimes occurs by freezing or mummification.

3. Replacement is the principal method by which fossils may be preserved. Body substance is replaced by more durable material such as calcite, silica, pyrite, etc.

4. Permineralization makes bone and wood heavier and stronger.

5. Distillation under high temperature may leave carbon residue film on a shale parting or bedding plane. This shows the outline of the fossil.

6. Indirect evidences provide information about an animal's presence and mode of life, even though the animal remains are not represented.

7. Sedimentary rock is the source bed of most fossils. They would have been melted in igneous rock and greatly deformed in metamorphic rock.

8. A marine environment, with its rapid burial and reducing conditions, preserves organic remains much better than a terrestial (land) environment. On land, organisms decay quicker, are oxidized in the air, and become torn apart and scattered by animals.

9. An index fossil identifies a formation and serves to locate it in the stratigraphic column. It may be used for correlation, as may also assemblages of fossils.

10. Exoskeletons are external covering or shells (for protection). Endoskeletons are internal frameworks (primarily for strength with mobility).

11. Genera are similar, but do not interbreed.

12–15. See characteristics in Additional Information.

Answers	
fossils	1
direct	2
replacement	3
permineralization	4
distillation	5
indirect	6
sedimentary	7
marine	8
index	9
exoskeleton	10
genera	11
protozoa	12
porifera	13
chordata	14
pisces	15

Characteristics used in taxonomy (classifying organisms) include: number of cells of the organism, number of cell layers, presence or absence of tissues or organs, type of symmetry, presence of segmentation, origin of coelom (body cavity), and type of embryonic cleavage.

CHARACTERISTICS OF LIFE are: obtaining energy, growth, reproduction, response to stimuli, and (often limited) motion.

Class Amphibia—lives on land or in the water, cold blooded, gills in young and lungs in adult, lays eggs which must remain in water until hatched, Devonian to present.

Class Reptilia—lives primarily on land, cold blooded, air breathing, lays eggs which do not need to stay in water, teeth all the same (and number in the thousands), first real development of legs, Pennsylvanian to present.

Class Aves—Birds, warm blooded, air-breathing, has feathers and wings, lays eggs, bills instead of teeth, Jurassic to present.

Class Mammalia—lives primarily on land, warm blooded, air breathing, has fur, young are born alive and cared for until grown, teeth differentiated, Cenozoic to present.

33 SOILS AND GROUND WATER
SELF-TEST

1. Of prime importance in soil formation is the _____ rock material.

2. The most common soil products of weathering of minerals are the _____.

3. After billions of years of weathering, the end result is the _____ grain.

4. The B-horizon is the zone of _____.

5. If a ribbon of soil can be formed easily and stays together, the soil is _____.

6. Soils with concentrations of iron oxides and clays in the B-horizon are termed _____.

7. A red soil color usually indicates the presence of an oxide of _____.

8. The last mineral to crystallize in a magma and also the last to weather is _____.

9. The top surface of the zone of saturation is called the water _____.

10. Contact between the top of the zone of saturation and the ground surface produces _____.

11. Contact between the base of the zone of saturation and magma heat produces _____.

12. The amount of connection between pore spaces is termed _____.

13. A porous and permeable bed is termed a(an) _____.

14. A tilted, confined aquifer (with a hydrostatic head) produces a(an) _____ situation.

15. A region of many caves and sinkholes has _____ topography.

16. A cave formation suspended from the ceiling is a(an) _____.

1 _____
2 _____
3 _____
4 _____
5 _____
6 _____
7 _____
8 _____
9 _____
10 _____
11 _____
12 _____
13 _____
14 _____
15 _____
16 _____

BASIC FACTS

ORIGINS OF SOILS. Soils are weathering products. They are the response (to new conditions at the surface) of minerals and rocks that were once in chemical equilibrium below the surface. The soil formed depends on the interweaving of several quite variable factors:

Soil formation factors include: original material, climate, relief position on the land surface, duration of exposure of land surface to the atmosphere, and type of vegetation.

Occurrence of Soils. The soil profile consists of several horizons or zones: *A-horizon* (leaching), *B-horizon* (accumulation), *C-horizon* (fragmentation), *mantle rock* (loose rock and soil), and *bedrock* (the solid rock beneath).

CLASSIFICATION OF SOILS. The classification of soils is built up of several factors: color of material (presence or absence of iron oxides or organic matter), consistency of moist material (loose, friable, firm), texture of material (clay, loam, sand, silt), chemical type (pedalfer, prarie, pedocal, laterite), permeability (slow, moderate, rapid), and absorption value (negligible, low, moderate, high).

Thus, a soil may be classified as a brown, friable, loamy pedalfer of moderate permeability and low absorption value.

SOIL REQUIREMENTS. The pH should be adjusted to meet the requirements of the crop to be grown, and to assure proper bacterial and fungous action. Bacteria require a pH of 6–8, and fungi 4–6. This is usually done by the addition of "lime." Tests should also be made to see if the addition of phosphorus, potassium, nitrogen, or organic material is needed.

(Continued on page 132)

ADDITIONAL INFORMATION

Weathering of the igneous feldspars or the sedimentary shales derived from them produces clay or silt. *Silts* are hydrous aluminum silicates of particle sizes from .02 to .002 mm diameter. *Clays* are hydrous aluminum silicates of less than .002 mm diameter. *Calcium* and the mineral calcite come primarily from the weathering of the calcic end of the plagioclase series. *Iron* (a common soil impurity) comes from weathering of the ferromagnesians and mica. These also produce the *magnesium*-laden ground waters that convert limestone to dolomite or deposit as magnesite (magnesium carbonate).

The insoluble quartz residue remains. It breaks down mechanically, ending as the sand grain.

Climate (particularly the amount of rainfall) drastically affects the soil type. *Relief* has an effect also as the soil may be different on a crest, a slope, or in bottom land.

The B-horizon contains concentrations of iron oxides and clays in humid climates and calcite in arid climates. These are primarily "leached" from the A-horizon above, which is gray or black with organic material.

Classification of soils is in terms of:
Soil colors—red = iron (semi-arid, arid); yellow = iron (subhumid, humid); black = organic matter; brown = iron + organic matter; gray = clay minerals.
Consistency—friable soil crumbles but sticks together.
Texture—some common field tests are: clay—forms a ribbon of soil easily and stays together; loam—ribbon doesn't form; sand loam—feels gritty; silt loam—feels smooth.

SOIL TYPES. *Pedalfers* have concentrations of iron oxides and clays in the B-horizon. They occur in temperate humid climates with moderate forest vegetation, as in the eastern half of the United States. *Pedocals* have concentrations of calcium carbonate in the B-horizon. They occur in temperate arid climates with sparse vegetation, as in the western half of the United States. *Prairie* soils are intermediate between these two. *Laterites* are red soils with hematite and bauxite concentrations in the B-horizon.

Meteoric water is water already in the hydrologic cycle. It is precipitated as rainfall, infiltrates, and is absorbed into the ground. *Juvenile water* is water that is new to the hydrologic cycle. It may be added by volcanic eruption. *Connate water* is water that has been removed from the hydrologic cycle, as for example, salt water found in drilling a well.

FEATURES OF THE WATER TABLE. The water table surface nearly parallels the top surface of the ground. The table rises under hills and is lower under valleys. The relief of the water table is greater in the rainy season than in times of drought. A *perched water table* is a local part of the water table underlain by an impervious bed that cuts it off from the

(Continued on page 132)

EXPLANATIONS

1. The original rock present is what basically determines the soil formed from it.

2. Clays are the weathering products of more different kinds of minerals than any other soil component.

3. The sands of the beaches of the world are the terminal deposit of billions of years.

4. The B-horizon is the zone of accumulation of the material leached out of the A-horizon.

5. Clay will easily form a coherent ribbon when rolled between the fingers.

6. Pedalfers (pedon, or soil, + Al + Fer) have concentrations of aluminum and iron in the B-horizon of this type of soil.

7. Red soils usually indicate the presence of iron oxides, such as the red hematite.

8. Quartz crystallizes last in a cooling magma and also is the last to weather since it is very insoluble.

9. In the zone of saturation any and all voids are filled with water.

10. The water table emerging at the surface of the ground produces springs.

11. A zone of saturation re-established over a volcanic region produces, by its contact with magma heat, hot springs.

12. Porosity is the amount of pore space present between grains (as in a sand pile).

13. A porous and permeable bed through which fluids will flow is called an *aquifer*.

14. An artesian situation may be an artesian flowing well, or even a spouting artesian fountain. It depends on the amount of hydrostatic pressure (head) and the resistance of the formation to flow of fluids through it.

15. *Karst* topography is the type of topographic surface displayed in an area where limestone outcrops at the surface in a humid climate. Here, there exist numerous sinkholes, caves, and underground drainage features.

16. Remember this mnemonic: "The stalactite holds tight to the ceiling."

Answers

original	1
clay	2
sand	3
accumulation	4
clay	5
pedalfers	6
iron	7
quartz	8
table	9
springs	10
hot springs	11
permeability	12
aquifer	13
artesian	14
Karst	15
stalactite	16

Origins of Soils

Igneous mineral	Sedimentary mineral	Soil component
feldspars	potassium-sodium-calcium clays	clays + soluble calcium
ferromagnesians	limonite, hematite, dolomite, magnesite	clays + iron
micas	clay and iron-magnesium ores	clays + magnesium clays + iron
quartz	quartz	sand

KINDS OF GROUND WATER: *meteoric* (absorbed rainwater), *juvenile* (new volcanic water), and *connate* (old buried seawater).

GROUND WATER ZONES: *zone of aeration* (air-filled crevices and pore spaces) and *zone of saturation* (water-filled crevices and pore spaces).

WATER TABLE. The water table is the top surface of the zone of saturation. Most of our drinking water comes from drilling into the water table. The height of the water table is directly proportional to the amount of rainfall—more rain, higher water table. It is nearly parallel with the ground surface.

Contact by the water table with the top surface of the ground produces: *springs* (natural contact) and *wells* (artifical contact). Contact with magma heat produces hot springs, mud pots, and geysers. *Recharge areas* of outrop exposure of aquifers (water-bearing formations) replenish the water table.

MOVEMENT OF GROUND WATER. Depends upon: *porosity* (amount of pore space present) and *permeability* (amount of connection between pore spaces).

WATER MOVEMENT STRUCTURES: *aquifer* (a porous and permeable water-bearing bed) and an *artesian situation*.

WORK OF GROUND WATER. *Solution* causes: sink holes, caves, swallow holes, rises, collapse tunnels, and Karst topography. *Deposition* causes: stalactites, stalagmites, columns, and helictiltes.

DEPOSITION WITHIN A FORMATION causes cementation and replacement.

main water table below. Heavy pumping of a well that bottoms within the water table will produce a *cone of depression*. A *spring* is natural, a *well* is man made. What the spring produces depends on conditions downslope from it. If the path is open, it flows as a stream. If blocked, a lake results. Choked with vegetation, this becomes a swamp. A swamp may also be produced where the water table lies at or very close to the surface of the ground. The crack, crevice, or tube of a *hot spring* is open. If the water passes through rock soluble in hot water, such as shale, *mud pots* are formed. If the tube is irregular and constricted (preventing temperature equalization by convection), a *geyser* is formed. The bottom water (under higher pressure) becomes superheated. Its eventual boiling raises the whole column (thus reducing the pressure) and it all turns to steam.

Movement of ground water demands both good porosity and permeability. The pore spaces may be filled, but water cannot flow unless interconnections exist between the pores and water can penetrate.

Sandstones (that are not tightly cemented) form good aquifers.

An *artesian situation* demands four things: an aquifer, surface outcropping at one end; an impervious bed above, preventing seepage up; an impervious bed below, preventing leakage down; a tilting of the bed to provide a pressure, where tapped, due to the vertical height of the water column (a hydrostatic head). With enough "head" pressure, water may reach the surface in a pipe as a flowing well. With more pressure, it may even rise as an artesian fountain. The Dakota Sandstone is the aquifer of the cities of the High Plains. It receives rainfall on the outcrops in the Rockies and is tapped hundreds of miles to the east and thousands of feet below the surface.

Solution takes place in limestone, first along joint cracks (which widen into funnel-shaped *sink holes*), and then along bedding planes which enlarge into different levels of a *cave* or *cavern*. Whole streams may disappear down these sink holes which are then called *swallow holes*. The drainage of an entire valley may go underground for 10 to 25 miles and then rise to the surface again along a joint crack. Caving in of the roof may cause intermittent *collapse channels* that outline the underground drainage course. A region with these features has *Karst topography* (named for an area in Yugoslavia).

Deposition forms include overhead *stalactites* and upward-building *stalagmites*. Where these meet, a *column* or *pillar* is formed. These cave formations grow about 1 inch in 100 years. Straw-like, very crooked stalagmites growing at random angles under a hydrostatic head are called *helictites*. The cave forms below the water table by solution and erosion by currents. Only when lowering of the table exposes the cavern to the air can we enter it.

34 ATMOSPHERIC LAYERS AND CIRCULATION

SELF-TEST

1. The zone of mixing at the base of the atmosphere is the _troposphere_

2. The zone of gas atom escape at the top of the atmosphere is the _exosphere_

3. The zone where atmospheric gases exist as ions is the _ionosphere_

4. The mesophere temperature rise is caused by ultraviolet absorption by _ozone_ gas.

5. Between the mesophere and the zone of mixing lies the _stratosphere_

6. Radio wave reflection occurs from the basic layers of the _ionosphere_

7. Charged particles are trapped in the geomagnetic field in the _Van Allen_ radiation belts.

8. Circumpolar, high-altitude light displays are _auroras_.

9. The present atmosphere is approximately __78__ % nitrogen.

10. The inert (uncombining) ingredient of the air is _nitrogen_

11. Heating of the air produces a _low_ pressure band at the equator.

12. The formation of three convection cells per hemisphere instead of one is caused by the earth's _tilt_.

13. The zones of winds between the equator and 30° (N and S) are the _trades_.

14. The high pressure bands at 30° (N and S) are called _horse latitudes_

15. The zones of winds between 30° (N and S) and 60° (N and S) are the _westerlies_

16. The "becalming" pressure band at the equator is the _doldrums_.

17. The zones of winds between 60° (N and S) and the poles are the polar _easterlies_

18. The deflection of moving winds from a straight N-S line is caused by the earth's _rotation_.

19. The force causing this deflection is called the _Coriolis_ force.

20. Deflection of a moving fluid is to the _right_ _____ in the Northern Hemisphere.

1 _____
2 _____
3 _____
4 _____
5 _____
6 _____
7 _____
8 _____
9 _____
10 _____
11 _____
12 _____
13 _____
14 _____
15 _____
16 _____
17 _____
18 _____
19 _____
20 _____

BASIC FACTS

ATMOSPHERIC LAYERS. The atmosphere may be divided into a number of zones on the basis of different physical and chemical properties. There are four different principal zones (from the ground upward): *troposphere, stratosphere, mesosphere* and *ionosphere*. The fact that they are determined by different properties and that other zones overlap them is not generally appreciated.

The four basic factors determining the zones of the atmosphere are: the *vertical extent of air turbulence* from the heating of the earth's surface, the *vertical temperature profile*, the gravity-layered *chemical composition* of the atmosphere, and the location and extent of the lines of force of the earth's magnetic field.

The present composition of the atmosphere (in a dry state) is *78% nitrogen* and *21% oxygen*, with the other gases about 1%. This is the remnant of an original atmosphere of methane, ammonia, and water vapor contributed by volcanoes. The molecular velocities of hydrogen and helium exceed, or come close to exceeding, the escape velocity of the earth at its surface. They have therefore withdrawn to layers farther out and colder, where they move slower, or have been lost to outer space. Carbon is tied up in rocks, organic matter, or carbon dioxide (as is most oxygen). Molecular nitrogen is inert and remains intact.

ATMOSPHERIC CIRCULATION. Circulation within the troposphere is a very complex, complicated, interweaving of wind and air currents. *Wind* is generally taken to mean air moving parallel to the ground. *Air currents* move in other directions.

(Continued on page 136)

ADDITIONAL INFORMATION

The *troposphere* (which extends about 7 mi up) may be considered the zone of mixing. Over 80% of the mass of the atmospheric gases are compressed into this zone. The unequal heating of equatorial and polar, land and oceanic, regions keeps these gases in constant motion. Almost all of the earth's water vapor (and consequently its clouds, precipitation, and storm disturbances) occur in this zone. The temperature lapse rate falls very evenly going upward (reaching about $-70°F$).

The *stratosphere* (7–15 mi) is so named because the air is in horizontal layers or strata with little vertical movement. The air here is cold ($-70°F$ to $-50°F$), clear, thin and calm.

TEMPERATURE ZONES. In the mesophere (15–50 mi), an interruption in the temperature decline rate is caused by the ozone layer which absorbs ultraviolet causing the temperature to briefly rise in this area. This absorption prevents the rays from burning life at the surface.

In the *thermosphere*, the velocity rate of gas molecules and atoms rises due to solar bombardment, hence the temperature response of these gases rises also.

CHEMICAL COMPOSITION. The atmosphere is gravity-separated, with the normal diatomic molecules making up the lower portion, and single atoms making up the upper.

In the *ionosphere* (50–250 mi), and above, most atoms exist as ions due to solar radiation stripping off electrons. The *exosphere* is so named because here some atoms exceed the escape velocity of the earth. This zone gradually grades into outer space.

ELECTROMAGNETISM. Note that the diatomic molecule zone is insulating, while the ionic zone is conducting. The *Van Allen radiation belts* consist of charged particles (protons and electrons) trapped in the earth's magnetic field. They are concentrated in two doughnut-like rings centered in the plane of the earth's magnetic equator. One is from 1000 to 3000 mi out, and the other is from 8500 to 12,500 mi out.

The *auroras* (aurora borealis, the northern lights, and aurora australis, the southern lights) are fluorescent gas displays caused by direct solar bombardment during sunspot activity, and also by radiation driven from the Van Allen belts at this time.

(Continued on page 136)

EXPLANATIONS

1. The troposphere is the lower air movement zone where most of our weather is caused.

2. The exosphere is the zone from which atoms escape when their velocity exceeds the escape velocity of the earth.

3. Ions form in the ionosphere as atoms are stripped of electrons by solar energy or particle bombardment.

4. Ozone (O_3) gas, which is the normal diatomic oxygen molecule with an extra oxygen atom, absorbs ultraviolet rays.

5. The stratosphere lies between the troposphere and the mesosphere.

6. Radio waves reflect from the E and F layers in the lower part of the ionosphere. Shorter waves reflect higher up. Ultra-short waves penetrate.

7. The Van Allen Radiation Belts are bands of charged particles trapped in the magnetic field of the earth.

8. Auroras center around the magnetic poles. Aurora borealis and aurora australis are called the northern and southern lights, respectively.

9. Our present atmosphere is about 78% nitrogen and 21% oxygen. All other elements make up less than 1% of the atmosphere.

10. Free nitrogen in the atmosphere does not normally combine with anything.

11. A low pressure band exists at the equator, due to heating and the rapid rise of air away from the surface.

12. The tilt of the earth's axis creates 3 cells per hemisphere, as the direct rays of the sun move alternately north and then south of the equator.

13. The trades lie between 0° (the equator) and 30° (N and S).

14. The horse latitudes are the 30° (N and S) high pressure bands.

15. The westerlies lie between 30° (N and S) and 60° (N and S).

16. The doldrums is a low pressure band at the equator.

17. The polar easterlies lie between 60° (N and S) and the poles.

18. The earth's rotation, or spin, causes deflection of moving winds from a straight N-S line.

19. The Coriolis force causes the deflection of moving fluids from a straight-line path.

20. A moving fluid deflects to the right in the Northern Hemisphere and to the left in the Southern Hemisphere due to the Coriolis force.

Answers

troposphere	1
exosphere	2
ionosphere	3
ozone	4
stratosphere	5
ionosphere	6
Van Allen	7
auroras	8
78	9
nitrogen	10
low	11
tilt	12
trades	13
horse latitudes	14
westerlies	15
doldrums	16
easterlies	17
rotation	18
coriolis	19
right	20

Four orders of wind features exist. They are (from largest to smallest): *first order* (tilt formed—permanent high and lows; modified by unequal distribution of land and water), *second order* (spin modified—trade, westerly, and polar easterly wind bands), *third order* (medium duration—transient cyclones (lows) and anticyclones (highs), and *fourth order* (short duration—local winds that vary from day to night; land and sea breezes, mountain and valley breezes).

A *simplified basic model* is easier to visualize, therefore it is conventional to take these orders up as isolated cases adding to a simple model. Thus the original model might be a nontilted, nonrotating, homogeneous-surfaced earth, with only one convection-pressure cell in each hemisphere. Heated air would rise at the equator and winds aloft would move to the poles.

COMPLEX MODEL. The *tilt* of the earth's axis changes the single cell to three, with winds moving from highs to lows. *Spin* deflects wind movement from straight N-S, by means of the Coriolis Force. Deflection is to the right in the Northern Hemisphere and to the left in the Southern Hemisphere.

The *simplest possible case* of atmospheric circulation would be the circulation patterns of an earth that was at right angles to the plane of the ecliptic (not tilted), not rotating, and all water (with no land surface). Such a model would have a single convection cell in each hemisphere, with a low pressure band at the equator and a high at each pole.

The effect of the earth's being tilted at its present $23\frac{1}{2}°$ angle is that the rays of the sun are directly overhead half of the year north of the equator and half of the year south of it. This breaks each hemisphere cell into *three cells* with the inner boundaries being around 30° and 60°. *At the equator*, a low pressure band exists where the air currents are moving up and away from the surface. This band of little lateral wind movement is called the *doldrums*.

At 30° (N and S), high pressure bands exist at the surface. Here also, very little lateral wind movement occurs at the surface. These bands were called the *horse latitudes*.

At 60° (N and S), low pressure bands exist at the surface of the earth.

At the poles (N and S), high pressure areas exist at the surface.

The *effect of the earths rotation* on these broad bands of circulating winds is *wind deflection*. Winds blow from high pressure areas to low pressure areas. Thus, the winds at 30° N would tend to blow straight north and south. The linear velocities, though, are greatest towards the equator.

A wind heading south from 30° is traveling slower than the point it is heading for, so it ends up behind it. When this is drawn on the globe, the vector is from northeast to southwest. Heading north, the wind outruns its point, and so it moves from southwest to northeast.

Winds are generally named for the direction they come from. The polar *easterlies* come from a general easterly direction. The *westerlies* come from a general westerly direction.

35   WEATHER—HUMIDITY AND CLOUDS

SELF-TEST

1. Atmospheric conditions at a particular time are called _weather_.

2. Cumulative atmospheric effects over a long time are called _climate_.

3. Wet- and dry-bulb thermometers are used in the sling _____.

4. The total amount of water vapor air might hold at a given temperature is called _saturation_.

5. The amount actually held is _relative_ humidity.

6. The amount held expressed as a percentage of what might be held, at a given temperature, is _absolute_ humidity.

7. An instrument using twisted fiber to measure humidity is a(an) _hygrometer_.

8. The temperature at which precipitation will occur is called the _dew pt_.

9. A barometer based on an evacuated cylinder is called a(an) _aneroid_ barometer.

10. Normal sea-level atmospheric pressure is _14.7#/sq"_.

11. This supports a mercury column that is _29.92_ inches high.

12. Spherical clouds are of the ~~cumulus~~ _cumuliform_ type.

13. Blanket-like clouds are of the _stratiform_ type.

14. Very high, wispy, feathery clouds of ice crystals are _cirrus_ clouds.

15. Medium-height, blanket-like, water vapor clouds that blur the images of the sun or moon are _altostratus_ clouds.

16. Very high, blanket-like, ice-crystal clouds that form halos around the sun or moon are _cirrostratus_ clouds.

17. Shallow ground fogs that "burn off" rapidly in daylight are _radiation_ fogs.

18. Fogs resulting from lateral transfer of horizontal air masses are _advection_ fogs.

19. Smog is sometimes caused by temperature _inversion_.

1 _____
2 _____
3 _____
4 _____
5 _____
6 _____
7 _____
8 _____
9 _____
10 _____
11 _____
12 _____
13 _____
14 _____
15 _____
16 _____
17 _____
18 _____
19 _____

Weather—Humidity and Clouds

BASIC FACTS

WEATHER AND CLIMATE. *Weather* refers to the atmospheric conditions at a particular time. It is a short-term picture of a combination of all the weather elements. *Climate* refers to the cumulative effects of atmospheric conditions over a long period of time. It is a long-term evaluation of the range of the weather elements. We speak of today's weather, or tomorrow's weather; but we speak of the climate of the last 50 or 100 years.

WEATHER ELEMENTS. Atmospheric conditions are determined by combinations of the weather elements. These are: *temperature, humidity, pressure, cloudiness, precipitation,* and *wind*. The weather elements serve to redistribute the uneven distributions of heat and moisture over the earth's surface. This is important to man as it greatly increases the amount of habitable land area.

Atmospheric temperatures are measured in Fahrenheit or Centigrade scales. *Humidity* (or the amount of moisture in the air) is usually expressed as absolute or relative humidity. It is measured by hygrometer or psychrometer.

Pressure is used rather than density since the composition of the fluid (at least in the troposphere) is assumed to be homogeneous. It is expressed in lb/sq ft, in bars or millibars (a bar is 1 million dynes/sq cm) or by in. or cm of mercury supported. It is measured by a mercury or aneroid (vacuum) barometer. The latter may be graduated in feet or meters as an altimeter (altitude meter).

CLOUD OCCURRENCE is the result of the condensation of very fine droplets (under 0.02 mm) of water vapor in the

(Continued on page 140)

ADDITIONAL INFORMATION

HUMIDITY AND DEW POINT are measured by hygrometer or psychrometer. The latter has two thermometers. One has attached to its bulb a cloth sleeve that is kept wet. This thermometer is therefore cooled by evaporation and always has a lower temperature. A rapid flow of air for rapid evaporation may be provided by a fan or by using a *sling psychrometer*, which has a swiveled handle allowing it to be swung around rapidly by hand. A *hygrometer* uses the twisting or untwisting of a fiber under tension to move a needle over a dial graduated to relative humidity. Wet bulb depression (dry bulb-wet bulb) is plotted against dry bulb temperature, giving relative humidity on one chart and dew point on another.

Saturation is the total amount of water vapor the air can hold at a given temperature. *Absolute humidity* is the amount actually held at a given temperature. *Relative humidity* (percent saturation) is the percentage of water vapor the air actually holds, compared with what it might hold at that temperature. Thus 80% relative humidity means the air has 80% of the water vapor it might hold at that temperature. If the temperature changes, both relative and absolute humidity will change, as they are inversely proportional. At absolute humidity, the air is saturated and no more water will evaporate into it.

Dew-point is the temperature at which precipitation will occur. If the temperature is at 82° and the dew point is 80° a 2° fall in temperature will produce some form of precipitation. If the amount of moisture in the air changes, the dew point will change.

The *pressure* of the *atmosphere* at sea level at 45° N latitude is 14.7 lb/sq in. This equals 1013.2 millibars of pressure and supports a mercury column 760 mm, 76 cm, or 29.92 in. high. The modern, more durable, aneroid barometer has no breakable glass parts and works by the movement of a flexible diaphragm that covers one end of an evacuated can. Movement of this diaphragm is linked to a graduated dial.

Cirroform clouds are the highest clouds of horizontal development, as shown in the table. They are usually fair-weather clouds, but may precede *stratiform clouds*, which are warm-front poor-weather clouds. *Cumuliform clouds* are fair-weather clouds.

Cirrostratus clouds do not blur or obscure vision of sun, moon, etc., but may form halos around them of 22° diameter

(Continued on page 140)

EXPLANATIONS

1. Weather is the current atmospheric conditions.

2. Climate is the sum of atmospheric conditions over a long time.

3. The sling psychrometer uses a wet bulb and a dry bulb to measure humidity and dew point. The wet-bulb temperature is always lower as it is cooled by evaporation.

4. Saturation occurs when the air can absorb no more water vapor.

5. Absolute humidity is the amount of water vapor actually held at a given temperature.

6. Relative humidity is a ratio and is expressed as a percentage.

7. A hygrometer usually uses hair or, sometimes, a plastic fiber.

8. Dew point is the point at which precipitation will occur when the temperature falls to that level.

9. An aneroid barometer uses a vacuum cylinder instead of a mercury column.

10. The normal pressure at sea level is 14.7 lb/sq in.

11. This supports 29.2 in. (760 mm) of mercury.

12–16. By definition.

17. Radiation fogs originate from ground cooling by radiation. The warmer air near the ground is cooled below the dew point after the ground loses its heat.

18. Advection fogs result from lateral horizontal air mass transfer. These are the "pea-soup" type of fog.

19. Smog sometimes comes from an inversion of the normal air temperature sequence. This puts warm air on top of cold air, and prevents pollution products from rising.

Answers

weather	1
climate	2
psychrometer	3
saturation	4
absolute	5
relative	6
hygrometer	7
dew point	8
aneroid	9
14.7	10
29.92	11
cumuliform	12
stratiform	13
cirrus	14
altostratus	15
cirrostratus	16
radiation	17
advection	18
inversion	19

Clouds of Horizontal Development

Height Prefix	Cumuliform	Stratiform	Cirroform
cirro- (ice crystals) Above 20,000 ft	cirrocumulus	cirrostratus (forms halo)	cirrus (wispy)
alto- (water droplets) Above 6500 ft	altocumulus	altostratus (blurs vision)	
500 ft	cumulus	stratus fog	

atmosphere. Condensation as clouds or precipitation occurs when the temperature reaches the *dew-point*. Humidity instruments measure dew-point as well as relative humidity.

CLOUD CLASSIFICATION is by direction of development and then by form and height.

Horizontal development by form: cumuliform (massive globular roughly-spherical clouds, sometimes with a flat base; associated with updrafts of warm moist air to higher levels), *stratiform* (sheet-like clouds of greater horizontal than vertical extent; may tend to cover the sky from one horizon to the other; associated with strong horizontal winds), and *cirroform* (wispy, feathery or tufted clouds; always high and usually scattered).

Horizontal development by height or altitude: high clouds (over 20,000 ft—have the prefix *cirro-* and are composed of ice crystals formed by sublimation below freezing) and *medium clouds* (20,000 ft to 6500 ft—have the prefix *alto-* and are composed of water droplets formed by condensation above freezing).

Vertical Development (clouds of greater height than width). Cumulus clouds are summer rain shower clouds. Cumulonimbus clouds are the larger summer thunderstorm clouds.

due to refraction by the 60° face of the ice crystal. *Altostratus* clouds do not form halos, but the water droplets obscure vision and blur the outline of the sun or moon. This cloud is the first one to produce rain in the oncoming sequence of the warm front. *Stratus* clouds usually cover the entire sky. They may resemble, or even be, fog or haze lifted from the ground.

Fog is a cloud low enough to the earth's surface to interfere with visibility. *Radiation fogs* occur when the radiative cooling of the ground lowers the temperature of moist air above it to the dew point, as on a clear night with light winds. These fogs first accumulate in low areas and are shallow ground fogs (visibility is better upwards than laterally). They dissipate or "burn off" rapidly in the daylight.

Advection fogs are the horizontal-transfer "pea-soupers" formed when air passes over a surface of different temperature and saturation. In the most common type, the air temperature is chilled to the dew point (warm moist air passing over a cooler surface). In another type, rapid upward evaporation occurs until saturation results (cold dry air passing over a warm moist surface).

Advection-radiation fogs are caused by daytime mass horizontal transfer, and by night-time radiative cooling. *Upslope fogs* form by adiabatic expansion and cooling (temperature change by expansion or contraction without transfer of heat by the air) as air is forced up a slope.

Smog (smoke + fog) may be caused by temperature inversion (warm air above cold air) which limits dissipation of noxious gases.

Cumulus clouds are the light, fluffy, puff-ball clouds seen in summer during fair weather, or behind the cold front.

The prefix *nimbo-* means rain and *nimbostratus* clouds are very low rain or snow clouds. *Fractostratus* clouds are ragged, broken, "scud cloud" patches, indicating considerable wind. *Stratocumulus* clouds are large, numerous, dense, individual, roll-shaped clouds. They are quite common in winter.

VERTICAL DEVELOPMENT CLOUDS. *Congested cumulus* clouds are large in vertical extent and massive. They have rounded tops and flat horizontal bases where the temperature reaches the dew point. These are our summer rain-shower clouds.

Cumulonimbus clouds are the highest and thickest of all clouds. These are the thunderheads associated with violent thunder and lightning storms. The top may reach a point where the rising warm air reaches the same temperature and density as the surrounding air. Here, it spreads out laterally. Under a strong wind at this level, it may assume a projecting anvil shape. Its top may be as high as 90,000 ft.

36 WEATHER—PRECIPITATION AND WINDS

SELF-TEST

1. Dew, frost, rain, snow, sleet, glaze, and hail are types of _precipitation_.

2. Conversion from water vapor to solid ice crystals below freezing at the surface produces _frost_.

3. Condensation above freezing above the surface produces _rain_.

4. Water droplets frozen when striking colder ground objects produce _glaze_.

5. Lines connecting points of equal atmospheric pressure are termed _isobars_.

6. When equal pressure lines are close-spaced the pressure gradient is _steep_.

7. An ideal wind, aloft, between straight parallel isobars is a(an) _____ wind.

8. A wind, aloft, between curved parallel isobars is a(an) _gradient_ wind. Surface winds below 3000 ft are modified from these by _frictions_.

9. The direction the wind comes from is called the _windward_ direction.

10. The direction the wind moves toward is called the _leeward_ direction.

11. Wind velocity is measured at the surface by the revolving-cup _anemometer_.

12. If leaves rustle and the wind is barely felt on the face, the wind is described as a _light breeze_.

13. If large branches of trees move, the wind is described as a _____.

14. If the roof of a house is severely damaged, but the house doesn't blow down, the winds are described as _strong_ winds.

15. High-velocity, high-altitude circumpolar ribbons of air are called _jet streams_.

16. The most sophisticated weather information gathering system uses data and photos from a(an) _orbiting satellite_.

1 _____
2 _____
3 _____
4 _____
5 _____
6 _____
7 _____
8 _____
9 _____
10 _____
11 _____
12 _____
13 _____
14 _____
15 _____
16 _____

BASIC FACTS

The *precipitation influence factors* that determine what will occur when the temperature reaches the dew point are: temperature, location (above the surface or at the surface), and type of change of state (condensation or sublimation). The latter occurs at lower temperatures.

The *precipitation types* are: dew, frost, rain, snow, sleet, glaze, and hail.

ISOBARS. Atmospheric pressure is shown by a map device similar to contour lines. Isobars are lines (on weather maps) connecting points of equal atmospheric pressure. Every point on a particular isobar line has the same pressure. The measurements are usually given in millibars.

PRESSURE GRADIENT. The wind blows from points of high pressure to points of low pressure. Thus, a *pressure gradient arrow* can be drawn showing the ideal direction of wind movement from a high-pressure isobar to a low-pressure isobar. The gradient (or slope) is always measured across the isobars at right angles to them. As with contour lines, when the isobars are widely spaced, the gradient is low (wind speed slowest); when the isobars are closely spaced, the gradient is high (wind speed greatest).

WIND TYPES. A *geostrophic wind* is an ideal wind blowing aloft at a steady rate between straight parallel isobars. This is caused by the force of the Coriolis effect which opposes the pressure gradient force, turning the wind direction from across the isobars to moving parallel to them. The force of the Coriolis effect becomes greater with increased wind speed and both reach a maximum when the wind is parallel to the isobars. At this point, the Coriolis force just balances the pressure gradient force and keeps the wind parallel. Isobars, however,

(Continued on page 144)

ADDITIONAL INFORMATION

PRECIPITATION TYPES. Precipitation is measured by the volume-graduated rain or snow guage. The major types of precipitation of water vapor are: dew (condensation above freezing at the surface), frost (direct freezing of water vapor below the freezing point at the surface), rain (condensation above freezing above the surface), snow (direct freezing of water vapor above the surface), sleet (water droplets frozen when passing through cold air), glaze (water droplets frozen when striking colder ground objects), and hail (pellets of ice and snow that have grown by repeated or prolonged passage through cold air).

Sleet is already frozen by the time it strikes the ground, glaze is not. On a cold ground object, glaze forms ice. Dew occurs when the radiative cooling of the ground lowers the temperature of the moist air above it to the dew point; at such times, light winds will produce fog. Repeated passage of precipitation through updrafts and downdrafts is the common cause of hail.

Wind direction is determined, at the ground surface, by the familiar *weather vane* and registered in compass direction or degrees azimuth. Winds are named for the direction from which they blow. (For example, a westerly wind comes from the west.)

Descriptive wind terms are: *windward* (direction wind comes from), *leeward* (direction wind moves toward—downwind), *veering winds* (revolve clockwise around a center), and *backing winds* (revolve counterclockwise around a center).

Beaufort Wind Scale (Modified for Land)

Description	Observation	Velocity (mi/hr)	Beaufort Number
Calm	Smoke rises straight up.	0–1	0
Light air	Smoke drifts.	1–3	1
Light breeze	Leaves rustle, wind felt.	4–7	2
Gentle breeze	Leaves and dust move.	8–12	3
Moderate breeze	Small branches and paper move.	13–18	4
Fresh breeze	Small trees sway.	19–24	5
Strong breeze	Large branches move.	25–31	6
Moderate gale	Whole trees move.	32–38	7
Fresh gale	Twigs break off.	39–46	8
Strong gale	Branches and roofing striped.	47–54	9
Whole gale	Trees uprooted, building damage.	55–63	10
Storm	Severe building damage.	64–75	11
Hurricane	Houses blown down, devastation.	75–up	12

(Continued on page 144)

EXPLANATIONS

1. The types of precipitation all involve water droplets or ice particles.

2. Frost is produced by direct conversion of water vapor to solid ice, without the liquid state in between.

3. Rain simply involves condensation from gas to liquid.

4. Glaze is water while falling. It does not turn to ice until it hits the ground.

5. Literally, the term isobar means equal weight.

6. A steep, or high, grade (gradient) gives rise to close-spaced isobars.

7. The geostrophic wind blows between straight parallel isobars.

8. A gradient wind between curved parallel isobars lies on the pressure grade or slope. Friction with the ground surface modifies surface winds from original geostrophic or gradient winds.

9. Literally, toward the wind.

10. Leeward is the direction the wind moves toward; it is the sheltered side.

11. The revolving cup anemometer is normally used to measure wind velocity at the surface.

12. A light breeze will rustle leaves and produce a wind that will just barely be gently felt on the face.

13. In a strong breeze, large branches move and the wind whistles as it moves around edges or wires.

14. If severe damage occurs, but not total destruction, storm winds are found.

15. Jet streams are circumpolar, high-velocity, high-altitude ribbons of eastward-moving air.

16. Orbiting satellites provide our most modern, sophisticated, complex, wide coverage, weather information data-gathering system.

Answers

precipitation	1
frost	2
rain	3
glaze	4
isobars	5
steep or high	6
geostrophic	7
gradient friction	8
windward	9
leeward	10
anemometer	11
light breeze	12
strong breeze	13
storm	14
jet streams	15
orbiting satellite	16

are usually not straight and often not parallel. Therefore, something less than the ideal geostrophic wind is usually found in nature.

More commonly found are curved isobars around high- or low-pressure centers. These tend to be more strictly parallel. They originate the second type of wind. A *gradient wind* is a wind blowing aloft at a steady rate between curved parallel isobars. This is caused by the interaction of three forces: the pressure gradient force, the Coriolis force, and centrifugal force. These winds encircle high and low pressure areas. Their direction of movement varies. Geostrophic winds and gradient winds are found at upper atmospheric levels.

Surface winds are found below 3000 ft, where friction with the surface of the ground affects their speed and direction of movement. Over a flat, level surface, deflection is about 25°. Over a rough, hilly surface, it may reach 45°.

Jet streams are high-velocity, circumpolar, tubular streams of winds traveling eastward at high altitudes. They occur between 35,000 and 40,000 ft, at the top of the troposphere, where warm, humid tropical air laps over cold, dry polar air.

The jet streams are about 300 miles wide, about 5 miles thick, and have speeds between 100 and 300 mi/hr and are useful "tail winds" for planes. The shifting patterns of the jet streams, and the resulting upper-air waves, determine the formation of the intermittent large-scale low- and high-pressure areas that dominate the temperate zone.

HYDROLOGIC CYCLE. The sea is the source of all atmospheric moisture. It moves from there through the hydrologic cycle of evaporation (humidity), condensation (clouds), precipitation, and run-off, and back to the sea. Some soaks in as ground water on the way.

Wind Velocity is measured, *at the surface*, by the revolving-cup *anemometer*, which is linked to a dial calibrated in mi/hr or m/hr. The Beaufort Wind Scale, developed by British Admiral Beaufort in 1806 for sailing ships, is widely used.

Aloft, weather element information comes from the combined output of various weather data-gathering systems such as: direct visual tracking with instruments of released balloon (wind direction, altitude and speed), tracking with radar of released balloon, data transmission by balloon-borne radio transmitter (air speed, movement, direction, temperature, altitude, pressure, and humidity), recording by rocket- or airplane-borne instruments, and transmission of satellite photos showing weather conditions.

The first two systems merely give wind information. The rest give much more. Each system gives information from a higher altitude and/or wider area then the preceding system. However, the cost increases in this order, also.

37 WEATHER SYSTEMS
SELF-TEST

1. M T is the symbol for a(an) _____ air mass.

2. C P is the symbol for a(an) _____ air mass.

3. The climatic zone at the North Pole is the _____ Zone.

4. A boundary surface between two air masses of unlike properties is a(an) _front_ _____ .

5. When warm air displaces cold air, there is a(an) _warm_ front.

6. A front is named for the temperature of the air that is doing the _displacing_.

7. A cold front overtakes a warm front. This creates a(an) _occluded_ front.

8. A gently sloping straight front, up which one air mass is thrust, is a(an) _____ . _warm front_

9. A steeply curved front where one air mass is burrowing under another is a(an) _cold front_.

10. Cumuliform clouds, light winds, and little precipitation are associated with _____ . _highs_

11. Stratiform clouds, strong winds, and rain or snow are associated with _lows_ .

12. The wind direction of an anticyclone is ~~~~ _outward_.

13. The wind direction of a cyclone is _inward_.

14. A wind system that makes seasonal changes from land to sea over a large area is called a(an) _monsoon_ system.

15. The direction a tornado moves toward is to the _north east_.

16. Tornadoes occur along _squall lines_ ahead of fast-moving cold fronts.

1. _____
2. _____
3. _____
4. _____
5. _____
6. _____
7. _____
8. _____
9. _____
10. _____
11. _____
12. _____
13. _____
14. _____
15. _____
16. _____

BASIC FACTS

The formation of (third order) *wave cyclones* stems from irregular continually-forming waves or loops formed by the upper air masses and the jet streams at their overlapping boundaries. These wave cyclones are sub-continental.

Local (fourth order) *weather systems* include: land and sea breezes, mountain and valley breezes, monsoons, and tornadoes.

AIR MASSES. International classification of air masses is based on three factors: *nature of the surface* of the source region (Maritime, M; Continental, C), *climatic location* of the *source* region (Artic, A; Antarctic, AA; Polar, P; Tropical, T: Equatorial, E), and whether the temperature of the air mass is warmer (w) or colder (k) than the *surface it passes over*. Thus, MTw is a maritime tropical air mass that is warmer than the surface it passes over.

CONVENTIONAL CLIMATE ZONES. *Arctic (Antarctic)*—65° to 90° N and S; *Polar*—50° to 65° N and S; *Temperate*—35° to 50° N and S; *Subtropical*—30° to 35° N and S; *Tropical*—15° to 30° N and S; *Equatorial*—0° to 15° N and S. Because of the earth's tilt, the Temperate Zone has an alternation of seasons. The Polar, Equatorial, and Tropical Zones are sources of air waves.

WEATHER FRONTS. A weather front is the boundary surface between two air masses whose factors (properties) are not alike. The formation of these fronts stems from one basic fact: warm fluids rise, cold fluids sink. Three basic types exist: *warm fronts* (where warm air is being thrust up over cooler air), *cold fronts* (where cold air is burrowing under warmer), and *occluded fronts* (where a cold front overtakes a warm front and one

(Continued on page 148)

ADDITIONAL INFORMATION

Climatic changes are usually more noticeable between one area and another than from one day to another. Climate classifications, therefore, usually are listings of regions of similar weather element averages. The conventional climate zones (shown in the Basic Facts column) are latitude-bounded and correlate with temperature. Other factors, such as pressure and humidity, may vary widely within a single zone, however. A better system may be to make a chart plotting along one side the three first order primary cell zones derived from the tilt of the earth's axis. These primary cell zones correlate with the second order wind bands caused by the earth's spin and with air mass source regions.

A weather front is the entire surface. A frontal line is the trace of its passage upon the ground. Most often, just the term front is used for both.

Climate Zones

Primary Cell Zones	Vegetation Zones			
	Forest	Low plain	High plain	Desert
Arctic 50°–90° N and S	conifer (fir, spruce)	tundra	antarctic	icecap
Temperate 30°–50° N and S	broadleaf conifer (pine)	prairie	steppe	shrub
Equatorial 0°–30° N and S	scrub forest	savanna	steppe	shrub

The surface of a *warm front* is a gentle slope (under 45°) up which warm air is being thrust over cold air (similar to the geologic thrust fault). Warm fronts move in a northeasterly direction. The beginning of the associated cloud sequence may be 1000 miles (2 days) ahead of the surface frontal line and over 20,000 feet high up the slope. Stable warm air (warmer than the ground) produces the whole range of stratus clouds along the slope, with precipitation for 200 miles out in front. If the underlying cold air is below freezing, this precipitation may be wet snow or sleet. The front may be up to 500 miles wide. It travels slowly (about 15 mi/hr) and thus brings several days of increasing cloudiness and precipitation. Behind it follow cumuliform clouds, rising temperature and pressure, and clear weather. Unstable warm air produces turbulence, cumulonimbus clouds, and thunder storms in place of the nimbostratus clouds and steady rain.

(Continued on page 148)

EXPLANATIONS

1. M = *maritime* (sea); T = *tropical*.

2. C = *continental* (land); P = *polar*.

3. The Arctic Zone is the climatic zone at the North Pole. In climatic terms, the Polar Zones (50 to 65° N and S) are nearer the equator than the Arctic Zones (65 to 90° N and S). Thus, the geographic poles (of rotation) are in the Arctic (Antarctic) Zones of climate rather than in the Polar Zones.

4. The front is the boundary wall between two dissimilar air masses. It is the entire boundary, not merely where it meets the ground.

5. In a warm front, warm air displaces cold air (moves it back out of the way and takes its place).

6. A front takes the adjectival part of its name from the relative temperature of the air mass that is doing the displacing.

7. An occluded front occurs where a faster-moving cold front overtakes a warm front. One of the two (usually the warm front) is raised aloft.

8. A warm front has a gently inclined straight slope (0°–45°), where warm air rides up over cold air.

9. A cold front has a steeply-curved convex surface where cold air burrows under warm air.

10. Highs are associated with clear weather.

11. Lows are associated with poor or foul weather.

12. The winds move outward from a central zone of high pressure called a high in an anticyclone.

13. The winds move inward towards a central zone of low pressure called a low in a cyclone.

14. A *monsoon* system is a seasonally reversible, large scale, land-to-sea wind system.

15. A tornado moves toward the northeast from the southwest.

16. Tornadoes occur along squall lines ahead of fast-moving cold fronts. Slow-moving cold fronts or warm fronts are not that violent.

Answers

maritime tropical	1
continental polar	2
Arctic	3
front	4
warm	5
displacing	6
occluded	7
warm front	8
cold front	9
highs	10
lows	11
outward	12
inward	13
monsoon	14
northeast	15
squall lines	16

of them, usually the warm front, is raised off the ground).

The air masses form high pressure areas. The fronts occupy low pressure areas between. The barometer falls when a front moves in; later it rises. Surface winds veer (shift) clockwise as a front passes (Northern Hemisphere). Cold fronts follow warm fronts in *frontal families*.

The first-order permanent highs and lows of the Polar and Temperate Zones are shifted semiannually by the seasons and the fact that part of the earth is a land surface. The Equatorial Zone maintains a rather permanent low pressure band.

From the semipermanent highs and lows of the extreme temperature zones, smaller air masses continually push into the temperate zone between. Their interactions create the *transient highs* (*anticyclones*) and *lows* (*cyclones*) that, by their never-ending procession, give this zone its variable weather.

Highs usually have cumuliform clouds, light winds, and little precipitation. *Lows* usually have stratiform clouds, strong winds, and rain or snow.

Tropical cyclones are called *hurricanes* in the Atlantic and *typhoons* in the Pacific. They are infrequent water convection eddies that are violently destructive storms.

Cyclone Families are successions of highs and lows that follow one another eastward across the globe.

The United States Weather Bureau issues daily weather reports and weather maps. Weather may be predicted by such factors as: assuming systems will continue to move at the same rate in the same direction and plotting barometric tendency to change in a given time.

Four orders of ocean circulation systems exists: first order—convection cells, second order—surface gyres, third order—deep countercurrents, and fourth order—local currents.

The surface of a *cold front* is a sharp convex curve, behind which cold air burrows in under warm air. Cold fronts move in a southeasterly direction. Two different kinds of cold fronts exist: slow-moving and fast-moving.

In a *slow-moving cold front* (20 mi/hr), associated clouds and precipitation form immediately in front of the upper forward edge of the curve and extend back over and through the cold air behind. Stable warm air in front of the curve will produce altostratus clouds overlying nimbostratus clouds and rain which penetrates the cold air behind. *Unstable warm air* will product turbulence, cumulonimbus clouds, and thunderstorms. The front may be 500 miles wide.

In a *fast-moving cold front*, the front moves rapidly because of fast winds aloft. A thin squall line precedes the front by 100–200 miles. It is characterized by cumulonimbus clouds, violent thunderstorms, and even tornadoes. The wind does not shift at this point, though, nor does the temperature change as it always does at the passage of any front.

An *occluded front* occurs when a faster moving cold front overtakes a warm front, and one of the fronts is raised aloft. The front that is raised would not show on a surface weather map but would show on a map of conditions aloft. Rain is produced accompanied by either altostratus or cirronimbus clouds depending on air stability. The occluded front moves eastward.

LOCAL SYSTEMS. A *monsoon* is a wind system that makes seasonal changes from land to sea over a large area. In January, for example, a high exists over Siberia and cool dry winds move south to a low over the Indian Ocean. In July, a low occupies Tibet and humid winds pour north bringing drenching rains to India from a high in the Indian Ocean to the south.

Tornadoes are extremely violent rotating wind funnels that originate from cumulonimbus clouds along a squall line. A partial vacuum exists in the center as the winds rush rapidly upward, and a building caught entirely in the funnel may explode outward from the normal air pressure inside. The funnel movement is always from the southwest. Wind speeds may reach 500 miles. The center of a tornado is filled with blue electrical activity.

Cyclones and tornadoes are two entirely different types of storms. The cyclone is very seldom a dangerous, destructive storm. In fact, most people do not even note its passage. The tornado is always dangerous and destructive.

Each surface current seems to have a deeper current beneath it of opposite temperature and direction. In fact, more than two such current layers may exist.

38

OCEANOGRAPHY

SELF-TEST

1. Water covers almost _____ of the earth's surface.

2. Water in the cold bottom zone of the ocean may be at 29° F because high _____ lowers the freezing point.

3. A line of sharp change from warm water above to cold water below is called a(an) _____.

4. Sea water contains dissolved gasses and metallic salts that are mostly _____.

5. Temperature and density control _____ circulation of water.

6. Terrigeneous sediments of the deep abyssal plains consist of _____.

7. Avalanche-like down-slope movements of loose, unconsolidated sediment, mud, and water are called _____ flows.

8. A destructive water wave originating from an undersea earthquake is called a(an) _____.

9. The wave movement is large but the particle movement is small in a(an) _____ wave.

10. The wave movement is small but the particle movement is large in a(an) _____ wave.

11. A water wave is actually a form of _____ energy.

12. A suddenly-increased wave height, due to reinforcement and resonance between wind and water waves, is called a storm _____.

13. Erosion of a steep irregular coastline produces, on the projecting masses, headland _____.

14. Erosion of a steep irregular coastline produces, in the indented parts, bay-shore _____.

15. A shallow body of water between an offshore island bar and the mainland is termed a(an) _____.

1 _____
2 _____
3 _____
4 _____
5 _____
6 _____
7 _____
8 _____
9 _____
10 _____
11 _____
12 _____
13 _____
14 _____
15 _____

BASIC FACTS

FIRST ORDER HYDROSPHERIC RELIEF FEATURES. *Continental-type topography* includes the continental shield and the continental platform. *Oceanic-type topography* includes the: continental slope, rise, abyssal plains, mid-oceanic ridges, and rift valleys.

SECOND ORDER RELIEF FEATURES: *sea mounts* (underwater volcanic mountain peaks), *guyots* (flat-topped sea mounts), *coral atolls* (reef structures fringing a volcanic core), *island arcs* (curving chains of islands, essentially volcanic, fringing the active coasts of continents), *trenches* (elongate areas of crustal downwarp that lie seaward of mountain arcs), and fracture zones.

Ocean floor sediments may be of *terrigenous* (land) *origin* or *pelagic* (floating) *origin*.

WATER MASSES, like air masses, have similar properties throughout. The contact surfaces between water masses are called *hydrospheric fronts*, or simply fronts. The horizontal layered nature of the water masses means that the fronts are usually horizontal rather than vertical.

Three vertical hydrospheric layers may be designated. These layers may be differentiated in terms of density, temperature variation, and penetration of solar radiation. The upper layer extends down between 330 and 660 feet. It is a zone of seasonal or geographic temperature fluctuation. The middle layer extends down to about 3300 feet. The lower layer is below 3300 feet.

CHEMICAL COMPOSITION. Dissolved salts (mostly chlorides) make up about $3\frac{1}{2}\%$ of sea water (sodium chloride is 2.3%). In addition, atmospheric gases are also dissolved in amounts varying by season and location.

(Continued on page 152)

ADDITIONAL INFORMATION

Almost 7/10ths of the surface area of the earth is covered by water (330,000,000 cu mi). If the land surface were entirely smooth, there would be over two miles of ocean covering it.

CHARACTERISTICS OF THE HYDROSPHERIC LAYERS. At a depth of one meter, the infrared (heat wave) portion (50%) of the spectrum has been absorbed, accounting for surface warming.

The top of the middle layer is called the *thermocline*—a line of sharp change from warm water above to cold water below. Here only blue and green wavelengths of visible light (3–4% of the spectrum) remain. This accounts for the blue-green color of deep water. Since the thermocline resembles a temperature inversion, vertical mixing is inhibited. The temperature declines slowly to near freezing at its base.

Cold (29°–40° F) water from the poles becomes denser, sinks, and collects in the bottom layer. The high salinity of this dense water lowers the freezing point of the water, allowing the low temperatures.

The terms tropomere, mesomere, and arctomere have been suggested for the upper, middle, and lower ocean layers, respectively.

Vertical Circulation is controlled by temperature and density. Density is affected by temperature and salinity. *Horizontal circulation* is controlled by wind and rotation. Surface polar cooling and tropic evaporation both reduce temperature and increase density. In addition, evaporation leaves behind salts which increase the density. Rainwater, however, lowers the density.

OCEAN FLOOR SEDIMENTS. The *calcareous ooze* consists of the shell remains of microscopic animals. After death, the shells drift to the bottom and build up at a rate of about one inch every 500 years. Slightly soluble in sea water, these shells are not found below a depth of three miles. Below this depth, insoluble siliceous shells of certain microorganisms are found.

Terrigenous (land-origin) Sediments. Shelf and rise sediments include gravel, sand, silt, and clay swept from the land as well as chemical carbonate precipitates. *Abyssal plain sediments* include red clays carried in suspenion that settle out here. *Turbidity currents* are long, narrow, avalanche-like masses of loose, unconsolidated, sediment-laden, muddy water. These move down-slope and account for the emplacement of much abyssal material.

The particle movement in a *water wave* is nearly circular, thus combining the parallel and right angle movements of the longitudinal and transverse waves. The *trochoid profile*

(Continued on page 152)

EXPLANATIONS

1. The surface of the Southern Hemisphere is 81% water, and that of the Northern Hemisphere is 61% water. The average is 71%.

2. High salinity may result from removal of pure water by freezing or evaporation, leaving the salt more concentrated.

3. Thermocline literally means "heat incline," but more often it is nearly horizontal rather than inclined.

4. Many of our most soluble metallic salts are chlorides. Thus, chlorides become concentrated in the sea.

5. The *vertical* water convection cells move on the sinking of colder, denser water and the rising of warmer, less dense water.

6. *Terrigeneous* literally means of the earth, as opposed to marine. Weathered clays are washed from the land into the ocean.

7. These huge tongue-like avalanches of muddy water are extremely turbulent. They have even broken transoceanic cables. Hence, the term turbidity flows.

8. A tsunami, or seismic sea wave, results from an underwater earthquake.

9. Progressive waves are the type that occur in an open body of water.

10. Standing waves are the type that occur in a restricted body of water.

11. The water wave is a form of moving kinetic energy. Principally, this is energy that is transferred from the wind.

12. A storm surge increases, may even double, wave height very suddenly by amplitude addition. (This is a simplified explanation.)

13. On a steep, irregular coast, headland cliffs form on projecting masses as wave erosion is concentrated on these points.

14. The material eroded from the headlands on this coast is carried into the bays and deposited as bay shore beaches.

15. A lagoon is a small body of water between a fringing bar or reef and the mainland.

ANSWERS

70%	1
salinity	2
thermocline	3
chlorides	4
vertical	5
red clays	6
turbidity	7
tsunami	8
progressive	9
standing	10
kinetic	11
surge	12
cliffs	13
beaches	14
lagoon	15

WATER WAVES, when not restricted, are complex progressive waves. In *progressive waves,* wave movement is large and free; particle movement is small. The wave is a form of kinetic energy. It normally represents energy transfer from wind to water. In the relatively fewer cases of marine earthquake disturbance, it represents transfer of seismic energy from rock to water. The height and steepness of waves are dependent upon the speed, duration, and fetch of the wind. (Fetch is the length of water surface the wind blows across.)

OCEAN CIRCULATION. Four orders of ocean circulation systems exist: first order convection cells, second order—surface gyres, third order—deep countercurrents, and fourth order—local currents.

WAVE EFFECTS ON A COAST. Wave action on *steeply sloping coasts* produces: wave-cut terraces, wave-built terraces, berms, foreshores, and back shores along the *shore*; and beaches, dunes, spits, hooks, tombolos, and stacks *inshore*. On *gently sloping coasts,* waves produce: offshore bars, breakers, and surf *offshore*; and lagoons, tidal inlets, mud flats, and salt marshes on the *shore*.

Either a tectonic change (land rises or lowers, sea doesn't) or a eustatic change (sea rises or lowers) may leave these shoreline features well above or below sea level.

SEA ICE. Three major kinds of sea ice exist: *pack ice* (formed by freezing of water in the open sea—thickness is usually under 50 ft), *shelf ice* (the seaward extension of continental glaciation—thickness is usually 1000 ft or more—found in the Antarctic), and *icebergs* (large masses of floating ice from continental or mountain glaciers entering the sea or from shelf ice).

of a water wave is what would be traced by an off-center point on a circular disk rolling along a flat surface.

Wave height increases with longer duration of the wind, longer fetch, and greater speed. Wave steepness (height divided by length) is great for a young storm, but decreases as height lessens and the length of the wave increases.

The movement of the water particle is not quite a circle. Successive orbits crest very slightly shoreward. This long-term creep is called *mass transport.* It is greater for steep waves. Winds and waves meeting a shore at an angle produce *longshore currents* and *beach drift*. These processes pile up water at the shoreline. Eventually this excess water escapes seaward at varying times and becomes powerful *rip currents* which expand offshore into *rip heads*.

On the lower parts of major rivers, flow alternates between fresh water out and sea water in. These portions are said to be *tidal rivers*. If very large, they are called *estuaries*. The incoming tidal wave, constricted in the river, often produces a 2–3 foot wall of water moving upstream. This is called a *tidal bore*.

When the wavelength of the water wave equals the wavelength of the wind, reinforcement and resonance occur. If they are in phase (which will depend on water depth), the amplitudes will become additive, giving a suddenly-increased wave height of perhaps 20 to 40 ft, called a *storm surge*.

COASTAL FEATURES. A *tombolo* is an island tied to the mainland by one or more sand bars. A *stack* is a pillar of rock cut off from the mainland by erosion.

Wave base is the depth to which the water is disturbed by waves. Below this, the water is calm.

Elevated shorelines show former levels of the sea (some may be 200 to 300 ft above present sea level). In contrast, now-sunken Mediterranean ports of ancient times and the dredging up of woolly mammoth skulls by fishing boats from the Grand Banks (150 miles east of the coast of Newfoundland) attest to former lower levels of the shoreline.

If the two are compared, *Arctic icebergs* are smaller and rough-surfaced, *Antarctic icebergs* are larger and flat-topped like the parent shelf-ice.

A special feature of pack ice is the *ice island*, a large flat-topped floating mass of ice enclosed within the arctic pack ice. Its thickness is over 50 ft, and it originates from a land glacier.

APPENDIXES

Appendix I—Solving Unit Problems

Most problem answers have two parts, a number part and a unit part, so most problems involve two steps in their solution.

Solving the unit step first will show whether the conversion factor selected was the right one. The proper conversion factor should have in the denominator of its unit fraction the same unit that was in the top part of the original unit fraction. Only then is cancellation possible. If the wrong conversion factor is used, the units will not cancel. Remember: You can only cancel from the top part of one fraction to the bottom part of another fraction.

Solving the number part is then mere manipulation.

ALWAYS WRITE BOTH COMPLETE NUMBERS AND UNITS.

Solving the unit step of a number-unit problem first will show whether the conversion factor used was the right one. Selecting a wrong conversion factor will throw the whole problem off, so this first step is a very important one. Another common source of error is omitting units within the problem as you work it out. If units are omitted, you cannot cancel and have to guess at the unit answer.

Sample problem: Velocity conversion. $350 \frac{mi}{hr}$ is how many $\frac{km}{hr}$?

Solution: Think of it in steps.

(a) Solve the unit step first.
Select right conversion factor.

$$\frac{mi}{hr} \times \frac{km}{mi}$$

Cancel common units.

$$\frac{\cancel{mi}}{hr} \times \frac{km}{\cancel{mi}} = \frac{km}{hr}$$

(b) Next, solve in terms of numbers. Insert the numbers and multiply.

$$350 \frac{\cancel{mi}}{hr} \times 1.61 \frac{km}{\cancel{mi}} = 563.5 \frac{km}{hr}.$$

Appendix II—Unit Systems

All parts of a problem must be in the same unit system. Three different systems are commonly in use.

	cgs System	mks System	fss System
Distance	centimeter	meter	foot
Mass	gram	kilogram	slug
Time	second	second	second
Force	dyne	newton	pound
$F = ma$	dynes = g × cm/sec²	newtons = kg × m/sec²	pounds = slugs × ft/sec²
$w = mg$	dynes = g × 9.8 cm/sec²	newtons = kg × 9.8 m/sec²	pounds = slugs × 32 ft/sec²
Work (Fd)	dyne-cm (erg)	newton-meter (joule)	ft-lb
Power (W/t)	dyne-cm/sec (ergs/sec)	newton-meter/sec (joules/sec)	ft-lb/sec

Note: kg is mass, lb is force (weight). This causes much confusion.

Different parts of a problem may be in different systems. The first step, then, is to change *all* parts to the same unit system. To do this use:

Conversion Factors

Mass: 1 slug = 14.594 kg (1 kg = 0.069 slugs); 1 slug = 32 lb downward force (weight).
Force: 1 newton = 100,000 dynes = 0.225 lb (1 lb = 4.448 newtons = 444,822 dynes).
Work: 1 joule = 10,000,000 ergs = 0.7376 ft-lb (1 ft-lb = 1.356 joules = 13,558,820 ergs).
Power: 1 watt = 10,000,000 ergs/sec = 0.7376 ft-lb/sec (1 ft-lb/sec = 1.356 watts).
1 horsepower = 550 ft-lb/sec = 746 joules/sec = 746 watts = 0.746 kilowatts
1 watt = 1 joule/sec = 0.00134 h.p.

Appendix III—Chemical Data

Acids and Salts: Formulas and Names

FORMULA OF ACID	NAME OF ACID			FORMULA OF SALT	NAME OF SALT			
	First Term		Second Term		First Term	Second Term		
	PREFIX		SUFFIX				PREFIX		SUFFIX
	PREFIX		SUFFIX				PREFIX		SUFFIX
HNO_2		nitr	ous	acid	$NaNO_2$	sodium		nitr	ite
HNO_3		nitr	ic	acid	$NaNO_3$	sodium		nitr	ate
H_2S	hydro	sulfur	ic	acid	Na_2S	sodium		sulf	ide
H_2SO_3		sulfur	ous	acid	Na_2SO_3	sodium		sulf	ite
H_2SO_4		sulfur	ic	acid	Na_2SO_4	sodium		sulf	ate
HCl	hydro	chlor	ic	acid	$NaCl$	sodium		chlor	ide
$HClO$	hypo	chlor	ous	acid	$NaClO$	sodium	hypo	chlor	ite
$HClO_2$		chlor	ous	acid	$NaClO_2$	sodium	per	chlor	ite
$HClO_3$		chlor	ic	acid	$NaClO_3$	sodium		chlor	ate
$HClO_4$	per	chlor	ic	acid	$NaClO_4$	sodium		chlor	ate

Oxides of Metals: Formulas and Names

	Valence Number		
+1	+2	+3	+4
Hg_2O mercur*ous* oxide	HgO mercur*ic* oxide		
Cu_2O cupr*ous* oxide	CuO cupr*ic* oxide		
	FeO ferr*ous* oxide	Fe_2O_3 ferr*ic* oxide	
	PbO plumb*ous* oxide		PbO_2 plumb*ic* oxide
	SnO stann*ous* oxide		SnO_2 stann*ic* oxide

Values of R for the Universal Gas Law: $PV = nRT$	
0.082	l-atm/mole-°K
1.987	cal/mole-°K
8.314	joules/mole-°K

Oxides of Nonmetals: Formulas and Names

FORMULA	NAME			FORMULA	NAME			
	1st Term		2nd Term		1st Term		2nd Term	
		Suffix	Prefix			Suffix	Prefix	
CO	Carbon		mon	oxide	N_2O	Nitr	ous		oxide
CO_2	Carbon		di	oxide	NO	Nitr	ic		oxide
SO_2	Sulfur		di	oxide	NO_2	Nitrogen		di	oxide
SO_3	Sulfur		tri	oxide	N_2O_3	Nitrogen		tri	oxide
P_2O_5	Phosphor	ous	pent	oxide	N_2O_5	Nitrogen		pent	oxide

Final Examination
Dictionary-Index

FINAL EXAMINATION

Questions

Mark each of the following questions true or false.

_____ 1
_____ 2
_____ 3
_____ 4
_____ 5
_____ 6
_____ 7
_____ 8
_____ 9
_____ 10
_____ 11
_____ 12
_____ 13
_____ 14
_____ 15
_____ 16
_____ 17
_____ 18
_____ 19
_____ 20
_____ 21
_____ 22
_____ 23
_____ 24
_____ 25
_____ 26
_____ 27
_____ 28
_____ 29
_____ 30

1. A kilogram is 100 grams.

2. A vector represents a directed force drawn to scale.

3. Ft-lb/sec is a unit of work.

4. The speed of a machine is directly proportional to its IMA.

5. Molecules of gases are constantly fixed in position.

6. The resistance of a liquid to internal flow is its viscosity.

7. The distance between corresponding wave points is the amplitude of the wave.

8. Electromagnetic waves travel best in solids.

9. A convex lens is thicker in the middle.

10. Angle of reflection is greater than angle of incidence.

11. Unlike magnetic poles repel.

12. Static electricity is electricity in motion.

13. A generator basically produces a direct current.

14. The unit of electrical current flow is the volt.

15. The nucleon with a 1 amu mass and a positive charge is the neutron.

16. An atom that has gained or lost electrons is an ion.

17. The number of protons in the nucleus of an atom is the atom's atomic weight.

18. The Periodic Table lists elements by increasing atomic number.

19. Fluorine, chlorine, and bromine are members of the halogen group.

20. A base turns litmus red.

21. Acid + base → salt + water.

22. The amount of heat needed to raise one pound of water 1° F is one calorie.

23. The heat of vaporization of water is 540 cal/g.

24. According to Charles' Law, volume is inversely proportional to temperature.

25. The mole concept makes it possible to convert between chemical units.

26. An alcohol is the organic equivalent of an oxide.

27. An ester is the organic equivalent of a salt.

28. TNT is nitroglycerine soaked in sawdust or wood chips.

29. Einstein's General Theory of Relativity deals with uniform motion.

30. An alpha particle is the nucleus of a helium atom.

31. Beta (−) emission decreases atomic number by one.

32. Atomic fission is the combination of small atoms into larger ones.

33. Radioactivity is the emission of radio waves.

34. Kepler described planetary motion in terms of cycles and epicycles.

35. The density of the outer planets is greater than that of the inner planets.

36. The planet with rings is Jupiter.

37. Population II stars are older and move at a faster velocity than Population I stars.

38. The Hertzsprung-Russell diagram shows patterns of star growth.

39. Quasars emit little light.

40. Minerals are made up of different rocks.

41. The major economic minerals are mostly oxides or sulfides.

42. Land sections are subdivided into townships.

43. Erosion is a passive process.

44. A youthful stream carves a U-shaped valley.

45. Valley glaciers are often tributary to Piedmont glaciers.

46. The movement of solid rock is called diastrophism.

47. Melted rock below the surface of the ground is called lava.

48. The outer light-weight rock of the earth's crust is called sial.

49. The major cities of the world are mostly built on the continental shelves.

50. The outer core of the earth is solid iron and nickel.

51. Continental drift implies that the Pacific Ocean was once narrower.

52. A relative time column shows the specific ages of some formations.

53. The major divisions of the time scale are based on the abundance of certain fossils.

54. Invertebrate animals always stand on their heads.

55. Reptiles were predominant in the Cenzoic Era.

56. The B soil horizon is the zone of leaching.

57. The water table is the top of the zone of saturation.

58. The troposphere is the atmospheric layer at the earth's surface.

59. Weather is a short-term atmospheric characteristic.

60. Lines of equal atmospheric pressure are called isotherms.

61. Warm air displaces cold air in a cold front.

62. Anticyclones are high-pressure-centered rotating wind systems.

63. The Coriolis effect is the deflection in a fluid due to the earth's rotation.

64. An underwater downslope avalanche flow is a turbidity current.

65. The depth to which water is disturbed by waves is wave height.

66. A rocket is an example of a reaction engine.

Answers

#	
1	false
2	true
3	false
4	false
5	false
6	true
7	true
8	false
9	true
10	false
11	false
12	false
13	false
14	false
15	false
16	true
17	false
18	true
19	true
20	false
21	true
22	false
23	true
24	false
25	true
26	false
27	true
28	false
29	false
30	true

EXPLANATIONS

1. A kilogram is 1000 grams. (The prefix *kilo* means one thousand.)

2. Its orientation shows direction and its length shows amount.

3. Ft-lb/sec is a unit of power (work per unit of time).

4. The speed is inversely proportional to the IMA.

5. Gas molecules are in constant random motion.

6. Viscosity is, by definition, the internal resistance of a liquid to flow.

7. The distance may be crest-to-crest, trough-to-trough, etc.

8. Electromagnetic waves travel best in a vacuum.

9. Convex lenses are thick in the middle and thin at the edges. Concave lenses are the opposite.

10. The angle of reflection equals the angle of incidence.

11. Unlike magnetic poles attract. Like poles repel.

12. Current electricity is in motion. Static electricity is at rest.

13. A generator basically produces alternating current.

14. The unit of electrical current flow is the ampere. The volt is the unit of potential difference.

15. The positive charge is characteristic of a proton, not a neutron.

16. Positive (or negative) ions have lost (or gained) electrons.

17. The number of protons in the nucleus is the atomic number.

18. Atomic number is used today; Mendeleev used atomic weight.

19. Fluorine, chlorine, bromine, iodine (and radioactive astatine) are the halogens.

20. A base turns litmus blue; an acid turns it red.

21. This is a neutralization reaction.

22. One Btu raises 1 lb of water 1° F; 1 calorie raises 1 g of water 1° C—at standard conditions.

23. The heat of vaporization (of water) is the heat necessary to convert 1 g (of water) at 100° C to steam.

24. The volume of a gas is directly proportional to temperature.

25. The mole is the "interchange on the chemical expressway."

26. Alcohols (with OH radicals) are the organic equivalent of bases.

27. Esters are formed (along with water) by the action of an organic acid on an alcohol.

28. Dynamite is nitroglycerine-saturated sawdust; TNT is the nitration product of toluene.

29. The Special Theory deals with uniform motion. The General Theory deals with gravity and mass.

30. The helium nucleus has two neutrons and two protons.

31. Beta (−) emission increases atomic number by one.

32. Atomic fission involves the splitting of large atoms.

33. Radioactivity is (most simply) the emission of alpha and beta particles and gamma rays.

34. It was Ptolemy who used cycles and epicycles.

35. Inner planet densities are about 5.0; outer planets, about 1.425.

36. Saturn is the planet with rings.

37. Population I stars are found in the central disc of the galaxy; Population II stars are outside it.

38. The Hertzsprung-Russell diagram traces star development in its life cycle.

39. Quasars, recently discovered, are the brightests celestial objects.

40. Rocks are made up of minerals (one or several).

41. Oxides are weathering products; sulfides are vein fillings.

42. Townships are subdivided into 36 sections, each one mile square.

43. Erosion is active because materials are transported.

44. Youthful stream valleys are V-shaped; glacial valleys are U-shaped.

45. Piedmont glaciers form from merging valley glaciers.

46. Solid movement is diastrophism; movement of melted rock is vulcanism.

47. Melted rock is magma below ground; lava on the surface.

48. Sial rocks are the outer light-weight part of the crust. Sima rocks are the heavier inner portion.

49. The continental shelves are the underwater parts of the continental platform.

50. The outer core is thought, from seismic evidence, to be fluid.

51. The Atlantic Ocean has widened according to this theory.

52. An absolute time scale would give specific ages.

53. Life is thought to have evolved sometime in the Precambrian Era.

54. Invertebrate animals are without backbones.

55. Reptiles were predominant in the Mesozoic Era. The Cenzoic is the present era.

56. A-horizon is a leaching zone; B-horizon is an accumulation zone.

57. Below the water table, the soil or rock is saturated.

58. The troposphere is the zone of mixing, with winds and storms.

59. Weather is short-term; climate is long-term.

60. Isobars are equal-pressure lines on a map; isotherms are equal-temperature lines.

61. A front is named for the temperature of the displacing air mass.

62. Winds rotate clockwise outward from the central high.

63. Northern Hemisphere deflection is to the right; Southern Hemisphere deflection is to the left.

64. Turbidity current flow fills in the abyssal depths.

65. The depth to which matter is disturbed by waves is the wave base.

66. It runs by Newton's Third Law of Motion (action and reaction).

false	31
false	32
false	33
false	34
false	35
false	36
true	37
true	38
false	39
false	40
true	41
false	42
false	43
false	44
true	45
true	46
false	47
true	48
false	49
false	50
false	51
false	52
true	53
false	54
false	55
false	56
true	57
true	58
true	59
false	60
false	61
true	62
true	63
true	64
false	65
true	66